THE AGE OF SCIENCE

THE AGE OF SCIENCE

The Scientific World-view in the Nineteenth Century

David Knight

Basil Blackwell

© David Knight 1986

First published 1986
First published in paperback 1988

Basil Blackwell Ltd
108 Cowley Road, Oxford OX4 1JF, UK

Basil Blackwell Inc.
432 Park Avenue South, Suite 1503
New York, NY 10016, USA

British Library Cataloguing in Publication Data

Knight, David, *1936 Nov. 30–*
    The age of science : the scientific world-
    view in the nineteenth century.
    1. Science—History—19th century
    I. Title
    509'.034   Q125
    ISBN 0-631-15064-1
    ISBN 0-631-16176-7 Pbk

Library of Congress Cataloging in Publication Data

Knight, David, 1936–
    The age of science.

    Bibliography: p.
    Includes index.
    1. Science—History. 2. Knowledge, Theory of—
    History. I. Title.
    Q125.K55 1986   509'.034   86-8290
    ISBN 0-631-15064-1
    ISBN 0-631-16176-7 (pbk.)

Typeset in 11/13pt Sabon by Columns of Reading
Printed in Great Britain by T.J. Press Ltd., Padstow

Contents

To each eye, perhaps, the outlines of a given civilization present a different picture; and in treating of a civilization which is the mother of our own, and whose influence is still at work among us, it is unavoidable that individual judgement and feeling should tell every moment both on the writer and the reader. In the wide ocean upon which we venture, the possible ways and directions are many.

J. Burckhardt, *The Civilization of the Renaissance*

Since all that beat about in Nature's range
Or veer or vanish; why should'st thou remain
The only constant in a world of change
O yearning thought! that liv'st but in the brain.

S. T. Coleridge, 'Constancy to an Ideal Object'

Foreword

This book began with an interest in scientific argument in the nineteenth century, and turned into an essay on the development of the sciences and their relations with other human activities. Science and history were the great discoveries, in a sense, of the nineteenth century: Victorians turned them from hobbies into systematic professional activities bearing upon the world-view of all thinking people, and thus affecting social and religious activities. The history of science is therefore a powerful instrument for exploring the intellectual history of the age, which was the Age of Science in a way that times before and since were not. At last the sciences were reaching maturity and could bring, through steam power, fertilizers, dyes and electricity, the improvements in life promised two hundred years before by Francis Bacon. And yet science was still innocent, and belief in the steady progress of civilization possible. Scientism, the belief that all human life is to be understood through science and that science can give us the values we need, was urged by various thinkers, and represented a major challenge to traditional Christianity, although many prominent scientists remained more or less orthodox members of churches.

By the end of the century, the growth of educational systems had led to definite frontiers being erected between the various sciences, and to a world where those trained in science and in the humanities no longer shared a common culture. Pure and applied science were also distinguished, in a way they had not been earlier. We are looking at a time of increasing specialization, where people began to learn more and more about less and less. And yet the nineteenth century also saw some great generalizations, for example, the principle of conservation of energy and the theory of evolution by natural selection, which brought together all sorts of separate ideas and observations.

There had been men of genius in the sciences at all times, but the nineteenth century was the period when more ordinary folk could hope to make a career in science. By the end of the century these included some women, and some people in, for example, Japan and India. We need not be interested only in the outstanding; the exponential growth of science comes to mean something when we look also at more typical scientists. We also see thereby that science is one of many human activities, having its differences from, and similarities and relationships with, others.

Nineteenth-century science could also be fun. Lecturers aimed to entertain as well as instruct, and the magic lantern and scientific toys were effective ways of arousing interest. Gas lighting and cheap printing made evening classes and public libraries great institutions of the period; many efforts were made to make scientific knowledge attractive. Its history ought to be equally entertaining.

The historian has to select from the masses of material available, and construct a narrative; indeed perhaps to construct a past. Readers should never feel that they have met some definitive account which can never be superseded, because all our judgements and interests are a little different. Success for a writer of history means engagement with readers who are stimulated by the story to read and think more.

A new book, covering the whole history of science from Aristotle to DNA in sixteen brief biographies is R. Porter (ed.), *Man Masters Nature*, London, 1987. For science and politics, see S. Collini, D. Winch, and J. Burrow, *That Noble Science of Politics: A Study in Nineteenth-Century Intellectual History*, Cambridge, 1983. On the importance of religion, and for the view that all history is an instrument of doctrine of one kind or another, see M. Cowling, *Religion and Public Doctrine in Modern England*, Cambridge, I, 1980; II, 1985; III forthcoming. On science and fun, G. L'E. Turner, 'Scientific Toys,' *BJHS*, 20 (1987) 377–98.

List of Illustrations

Acknowledgements

I would like to thank the University of Durham for granting me sabbatical leave in the summer of 1984, when much of this book was written; and the publisher's reader, who made valuable comments on an earlier version. Figures 3–6, 9 and 10 were photographed by KNEST Photographic, Durham.

1

The Age of Science?

We think we live in the Age of Science. All around us we see its signs and wonders. Our civilization depends upon science and technology, which are closely related – though not perhaps as closely as is commonly believed. Ideas of what is real depend upon what is scientifically testable; and scientists are among the gurus of society. We believe that we live in a time of unprecedented change; and scientific investigation promises further change for the better, and claims to be the only really progressive human activity.

And yet we also see science as a threat: as dehumanizing through materialism, and more directly as the source of terrible weapons and of mechanisms of oppression. The days of 'logical positivism' are past, so that we no longer see scientific reasoning as the only true kind, or expect for example that ethics might be reduced to a science. Science itself is a fragmented activity: technology is not simply applied physics and chemistry, and between the various sciences there are frontiers which are not easy to cross. The training and methods of research in different sciences are different, and there are differences in status between practitioners of different branches: the border, for example, between chemistry and physics is a most interesting one, set about with nervous jokes.

The idea that there is a 'scientific method' in anything but the most banal sense has been generally abandoned: it is true that anybody working in the sciences must take account of observations and perhaps do experiments, but people work within research traditions or paradigms. For a time and within a particular field such paradigms generate theories and make problems interesting; over longer periods wider world-views change, making what had seemed central at one time now appear pseudo-science or at least a blind alley. We may not believe with Feyerabend that 'anything goes', or with Einstein that the scientist should be an unscrupulous opportunist where philosophical ideas are concerned; but the provisional nature of scientific thought cannot now be doubted. All this makes it clear that science is a human activity, not a royal road to truth; and being human is a matter of knowing in part. This should make science and its history more, not less, interesting.

The age of science is very great: it begins with megaliths, with Hindu mathematics, Egyptian and Chinese engineering, and Greek speculation and geometry. By the seventeenth century, with the addition of the Judaeo-Christian idea of man made in God's image and thus able to understand His work, modern science began in western Europe. Francis Bacon's vision of a world in which science produced plenty in a restoration of the Garden of Eden; Galileo's use of the telescope, and application of the physics of the Earth to the Moon and planets; and Descartes' new philosophy and mathematics marked a new way of looking at nature. Questions were to be asked which had definite answers; nature was put to the question like someone before the Inquisition. The invention of printing meant that science could become public knowledge: texts, and tables of figures, could be accurately reproduced and widely circulated, and work could be rapidly disseminated so that science became a cumulative affair, each man standing upon the shoulders of his predecessors. The founding of the first scientific journals, and of the first academies of sciences, in the second half of the seventeenth century gave institutional form to the new activity.

By 1700 science had become something to talk about, something which an educated man (and increasingly an educated woman) should know something about. For doctors, lawyers, clergy and landed gentry, it might be a vocation or a hobby; their interests might range across our whole spectrum of science, embracing perhaps physics and biology, along with antiquarian and literary pursuits. But to live by science was almost out of the question in the seventeenth and eighteenth centuries: a few, the analogues of the Grub-street hacks of literature perhaps, managed to live by writing and lecturing, but it was a precarious career. This was the prelude to the Age of Science; a time when scientific research was still simply recherché, the delight of dilettantes rather than the occupation of scientists. The history of these centuries can perhaps be written with little reference to science, as that of the first two centuries of the Roman Empire can ignore Christianity; except that unlike the early Christians, the early scientists (even Galileo) were generally courted by the powers that be.

Down to the end of the eighteenth century, science was a programme; a blank cheque drawn upon the future; an infant in the intellectual realm. It was during the nineteenth century that it came to maturity; the Age of Science was that century, interpreted perhaps in a liberal way as running from 1789 to 1914, the French Revolution to the Great War. The fall of the Bastille seemed a dawn in which it was bliss to be alive, but promises were unredeemed; to Humphry Davy lecturing in London (*Works*, II, 323) at the turn of the century science provided a contrast to that visionary gleam: 'We look for a bright day of which we already behold the dawn.' It was not political revolution but scientific development which would bring prosperity and reduce misery. This would happen both through science being applied to what were previously activities done by rule of thumb, traditional routines; and also by the general adoption of 'scientific' ways of thinking. This was the programme of the Age of Science; an age of innocence and of faith.

Christianity came of age when the Church was recognized under Constantine, and in due course became the established religion of the Roman Empire. Its institutions were then fixed into forms which persisted for centuries, while those who dissented from it were punished by the secular arm. T. H. Huxley a century ago described the scientific establishment as 'the Church Scientific'; by that time it had reached indeed something like that position. There were important posts in science whose holders formed something like a hierarchy: they held prestigious professorships, belonged to Academies, lectured and gave presidential addresses which were widely reported, and mingled with the mighty. The ambiguous role played by Constantine in the Christian Church was perhaps taken by Napoleon or

Bismarck, or by the less alarming figure of Prince Albert. T. H. Huxley preached what he called 'Lay Sermons', partly against organized religion; and although the scientific lecturer held forth at ground level like an actor to an audience raked around him, his performance was not altogether different from a preacher's in a pulpit above the heads of his congregation: he was indeed like a Wesleyan Sunday-school teacher.

During the Age of Science the boundaries between the sciences were set, much more stably than political boundaries have proved to be. The century was one of some great syntheses – the laws of thermodynamics, the theory of evolution – but also one of specialization: Adam Smith's principle of the division of labour was carried over into the intellectual world. The Age of Science was an age of sciences. In the eighteenth century the Renaissance ideal of the universal man was still almost possible; we think of Diderot, Franklin, Priestley and Goethe. There were still giants in later science; but John Herschel's decision not to specialize was uncommon by the 1830s, and would have been unwise thereafter. The future was with the expert, and also with the full-timer. By the end of the nineteenth century, a scientist would expect to read books and journals confined to his own specialism – perhaps inorganic chemistry; and to rely upon what was essentially popularization for all other parts of science.

With all this went a change in the nature and scope of scientific societies. At the beginning of the century they were general, covering all sciences: and since the word 'science' meant any organized body of knowledge, it was not inappropriate that the Royal Society, for example, should contain Fellows with only a genial and amateur interest in chemistry or geology. It was a club, where active researchers could meet friends and potential patrons, and would hear papers on all aspects of science. Able men who made good connections might expect like Davy to be elected in their early twenties, having got to the frontier of a science in a year or two of study, and begun research. By the end of the century, almost all scientific societies were learned ones, in that they were open only to the qualified. The Royal Society in Britain and Academies elsewhere had become a kind of body of elders, giving advice and administering grants. Journals like the Royal Society's, which had been general, now came out in sections covering only a part of the spectrum of science.

Nevertheless, the nineteenth century was the Age of Science partly because the 'two cultures' had not yet diverged. Those working in the sciences and those engaged in literature, painting, politics and theology all shared much common ground. Education and reading was less specialized, though by the end of the century this was rapidly changing. The science taught to those who were being trained in sciences was dogmatic and demanding, unlikely to provide metaphors or to excite the imagination. Science learned from a textbook, or taught then as part of a syllabus for a public examination, is rather different from that picked up from Mrs Marcet's *Conversations on Chemistry*[1] or from a demonstration-lecture by Faraday at the Royal Institution. To those who had picked it up less formally, as everybody did down to the 1870s, science was a seductive pattern of reasoning, which might be extended into other fields: geology was a newcomer, and why should not political economy or even painting join it?

This was the power of scientism, the idea that science is the guide to all reasoning and will provide the answers to all the questions which can reasonably be asked. Behind it lay the faith that the answers given in the sciences were independent of time and place; that they were truths, and that a scientific method led to certitude. In the early Church, the contemplative life had been looked upon as the ideal. The real Christian would spend a life of prayer on a pillar or in a monastery; weaker brethren and sisters would marry and do practical work, as estimable but second-class citizens. With the Reformation, monkish ideals were seen as immature; the whole man or woman should engage in all the practical problems of life, and not seek to opt out into some protected and sexless sphere in which he or she could be of no use. In this spirit, devoting one's life to the increase of knowledge was suspect; the well-balanced individual would not irresponsibly over-indulge himself in research.

There is a sense in which scientists, and indeed all intellectuals, are like children: they carry into adulthood their curiosity and argumentativeness, when most people by middle age are concerned with possessions, power and status: this is chiefly why it is a good idea to put them into universities, exposed to students, rather than research institutes where there is only academic politics to play. It is curious that in the Age of Science the contemplative life that is scientific research came to acquire a new respectability. In science as in the early Church, those who do this 'monkish' activity are the first-class citizens; the idea of the life spent in pure research rather than in practical activity is a fruit of the Age of Science, and is now under attack as monasticism was.

Naturally this pure research had practical effects, just as the monastic orders did; but the ethic of devoting one's life to the impartial search for truth was one of its great attractions. Science thus came to play a very important part in making the mould in which our society is cast. Ideas that were scientific, like Darwinism, or were believed to be, like Marxism and liberal political economy, dominate much of our thinking. Although science seemed to be above the vagaries of the day, in fact social imagery was extremely important in science, while scientific theory provided a metaphor for society. To try to understand the history of the nineteenth century without its science is a hopeless task, a struggle with one hand tied behind one's back; it was the Age of Science, and it is sad how neglected the science often is. To those interested in the history of ideas, it is the scientific ideas which in that century are the leading ones: other fields reacted to science. And in that century of newspapers, telegraphs, and education for all, the history of ideas is central.

It may be feared that compared to the science of Galileo and Harvey, that of the nineteenth century is arcane and inaccessible except to those who enjoy that kind of thing. Nothing could be further from the truth. It was the Age of Science because those engaged in the sciences took pains to make the world aware of their work and its implications. They wrote about it in essays in the *Edinburgh Review* and the *Quarterly Review* and similar general publications, they spoke about it at public meetings; it was popular science which was disseminated, and was picked up. Papers on thermodynamics, chemistry or physiology can indeed often be forbidding, and were so. The researcher in the history of science must go through them, as the researcher in science had to do; but such things have always been written for a small readership. Scientists of the last century managed to write about their science in an easy, even sprightly, way; and there is no reason why we may not also approach it in a spirit of respect tinged with irony.

Great rows were a feature of nineteenth-century science. Perhaps they are a sign of liveliness; of rivalry in an area where, as in the ancient Olympic Games, there is a prize only for the first past the post; of generation wars in rapidly changing disciplines. They are also a sign that even in its classical heyday science was never a sphere of certainty. Argument is a central feature of the sciences, and even the most abstruse papers have a rhetorical intent – they are written to persuade readers. Reputations are at stake, and the value of lifetimes of research may be called in question, when great challenges are made to the Church Scientific – as for example by Darwin and his allies. It gradually became clear to those inside the sciences that often a balance of probabilities rather than certainty was going to come out of such arguments; outsiders only perhaps became aware of this in our century. The hope in the Age of Science was that arguments would clear the air, that personal prejudices would be shown up, and that truth would emerge; to all it was evident that an experiment or an equation did not carry conviction on its own, but as part of a carefully worked-out explanation.

Looking back from our time to the frequently confused and unhappy nineteenth century, it is curious how many people seem to see only the confidence of 'Victorian Values'. This is a recognition of it as the Age of Science; for it was on the whole the scientists who were the optimists. They could afford to be. At last their pursuits were bearing fruit and achieving recognition; the future seemed to be theirs; everywhere they looked, they saw new worlds to conquer. Science became a profession in the later nineteenth century, just as it became a paradigm, an example, for all other forms of intellectual life: it was achieving its manifest destiny. To contemporaries who had grown up in rural societies, the change to an industrial society based on science and technology was daunting. They had more reason than we to feel that they were living through a period of unprecedented transformation; taken by scientists on a tram-ride down the 'ringing grooves of change' and yet, as Tennyson's poems remind us, left with sad longings.

Though it was the Age of Science, it should not be supposed that the scientists had it all their own way. Some of their arguments were with one another, others with those in fields which for one reason or another got left out of the magic circle of the sciences. Some of these we shall meet: phrenology and psychical research are important fields for the student of nineteenth-century science, rather as Freudian psychoanalysis will be to the historian of our days. To describe something as a pseudo-science is easy only with hindsight, and controversy is often seen best from the 'wrong' side. Phrenology, the science of bumps, aimed at being a kind of experimental psychology; it aroused excitement and indignation in many, seeming the latest thing in science, but failed to establish the connections between bumps and character, or even between brain development and bumps. The phrenologist in the end was a quack; he could not give satisfactory explanations.

Theologians talked about the 'scientific' part of their discipline, meaning the rather dry and formal parts of it. Huxley and others saw themselves as in competition with churchmen, but in general the Age of Science was not one of warfare of science and religion. There were skirmishes, but relations were more complex and more interesting, as will emerge. At the end of the century, the study of law, of languages, of history, of doctrine, of philosophy, was no longer generally spoken of as 'science'. There is no necessary reason why this should be so; in effect it marks the end of the Age of Science because it implies that ways of thinking which are clearly essential to society are in some way 'unscientific'. The sciences are a group of disciplines which in the 1830s fitted in with the programmes of the British Association for the Advancement of Science, and since the 1870s have formed Faculties of Science in universities; but they have no more of a natural frontier than the poor Poles.

Instead of expanding into new fields, the dogmatically trained scientists of the later nineteenth century allowed themselves to become a second culture. As a result both cultures were impoverished. The Age of Science saw the autonomy of the sciences; they were no longer under the control of anybody else. Their claims to other territory were, however, given up in an uneasy truce, not just with theology but with all arts and social sciences. Science without scientism is a rather narrow and specialized activity, with all sorts of interesting connections with our lives, but without the excitement of the last century; it is certainly not on its own something which can lead to a millenium. Maintaining civilization requires that we study also other things of equal or greater importance.

As we shall see, the nineteenth century became the Age of Science partly because in those years it became not merely an intellectual activity, but also a practical and a social one: an agent, as prophesied, in changing society. It was something done by groups and teams, with complex interrelations and sharing an education – generally in universities, which in the Age of Science became centres of learning rather than of teaching. Arthur Schuster in 1911 quotes an eminent Cambridge mathematician of the previous generation: if an undergraduate 'does not believe the statements of his tutor – probably a clergyman of mature knowledge, recognised ability, and blameless character – his suspicion is irrational, and manifests a want of power in appreciating evidence, a want fatal to his success in that branch of science which he is supposed to be cultivating'.[2] There were therefore those who saw no need for students to do experimental work, and saw their own proper role to be authoritarian teachers; but the new model was of a more critical discipline, and of everybody learning at different levels.

These different aspects of science brought it – indeed not unlike a church – into contact with all sorts of other activities in the nineteenth century. The story told here is not simply that of the progress of a triumphal chariot rushing towards our own day; rather it is an essay towards placing science in the cultural context of the nineteenth century, in the belief that it occupied a place there of paramount importance. Our story begins with France in Revolution and Empire, and with foreign reactions to the harvest of science there, from seeds fed and watered by these governments.

All history involves selection, and nobody can interest his readers in what he is not interested in himself, or is ignorant about. This is an essay, in an enormous sphere where a full treatise would be impossible. It may be that the choice of themes and characters is idiosyncratic; other historians might choose to give different emphases, and to pick quite different examples; or they might believe that in this field ranking and quantification are what is needed. But I hope that there are here shining instances, representative, typical and instructive, showing the growth of a scientific culture and making us from time to time turn away from the most eminent to the more ordinary, and from the genius to the popularizer.

2

Rivalry with the French

If anyone had asked, during the quarter-century of wars that followed the French Revolution of 1789, where true science was to be found, the answer would have been 'in France'. More strictly, for that country's intellectual life had become ever more centralized, 'in Paris'. Charles-Adolphe Wurtz, as a French patriot smarting from the defeats by the Prussians in 1870, declared that 'chemistry is a French science'. For most of its history this has not been true – if one had to place it somewhere, chemistry would be a German science – but in that Revolutionary and Napoleonic epoch it was true. French chemistry could even spare A. L. Lavoisier, executed in 1794 as a tax-collector of the *ancien régime*, for giants like C. L. Berthollet, A. F. Fourcroy and Guyton de Morveau survived and achieved high positions in the new regimes, bringing on the next generation.[1] In astronomy and mathematics the position was equally daunting to the true Briton. Just as the 'fixed air' of Joseph Black and the 'dephlogisticated air' of Joseph Priestley had been Frenchified into carbon dioxide and oxygen, so the mantle of Isaac Newton and Colin Maclaurin had passed to J. L. Lagrange and P. S. Laplace. Here too was a rising generation of exceedingly able mathematicians: manned flight also began in France.[2]

In natural history and comparative anatomy, the countrymen of William Harvey and John Ray had given up the search for a truly natural system of classifying organisms in favour of the convenient but artificial system of information-retrieval devised by C. Linnaeus in the middle years of the eighteenth century. A wealthy Englishman, Sir James Smith, bought Linnaeus' collections after his death, and in 1784 they came to London; whoever controlled the specimens in effect controlled the naming of plants and animals. But Linnaeus' convenient system relied on external characters, and a series of Frenchmen tried to work out something more fundamental and natural.[3] In 1789 Antoine de Jussieu published his *Genera Plantarum*, which has been the foundation of plant classification ever since; but for nearly forty years the British botanical establishment hesitated to take the plunge into these strange seas of thought. Gilbert White's *Natural History of Selborne* also appeared in 1789, and this classic indicates how wide the English Channel was: on the one side a Professor at a research institution with a museum and a great botanical garden, and on the other a country parson with a keen eye for wild life in the fields and woods. Charles Darwin was in some ways closer to White than to Jussieu.

When war broke out with the Revolutionary government and continued, with brief 'half-time' for the Peace of Amiens in 1802 and another break in 1814, until the Battle of Waterloo (1815), the British were cut off from much of the Continent. Scientific news travelled fast between France and Britain, but exchanges with places further away became much more uncertain; and the meeting with men of science from other countries and other traditions, which is so important in intellectual life, became almost impossible. The nation emerging as the chief defender of legitimacy against usurping regimes needed to see itself as not simply conservative. Philistine beef-eaters alone could not persuade Europe into constitutional monarchy on the British model; the spirit of Francis Bacon and of Newton must be rekindled amongst their descendants. The war must be fought on the intellectual as well as the military front; British science must become as good as French.

There was a further problem. Just as the early Fellows of the Royal Society in the 1660s had been anxious to show that their work had no associations with Thomas Hobbes, so men of science (or 'natural philosophers') of the 1790s had to demonstrate that there was no necessary connection between science and the overturn of established order. Scientists tend to be conservative, preferring a firm government that will let them get on with their work, and will provide reasonable patronage; and so it was in Britain, where there were few who cared much for the doctrines of the *philosophes*. Voltaire, Condorcet and Diderot may have indeed built upon Locke and the English Deists; but they found few disciples in the British scientific community. The most eminent scientist–politician was Priestley, whose support of the French (and of philosophical materialism and unitarianism) led to his being driven into exile in 1794: a fate certainly better than being beheaded, but which seems to show that a constitutional monarchy needed its chemists no more than a republic. It would be necessary to show that science need not have those political characteristics which had made Priestley unpopular with the powers that be, not only in England but also in America under John Adams: to carry on scientific research without, as Davy put it, yielding to 'the prejudices of the French'.

Because the revolutionaries had had the best of the propaganda battle, presenting all opponents as simply doing well out of the system, George Canning and others associated with William Pitt began in 1797 a new publication, the *Anti-Jacobin*, to expose the mistakes and lies, as they saw it, of left-wingers. One of the writers was John Hookham Frere, a diplomat and translator of Aristophanes, who helped make the magazine almost unique amongst pro-government propaganda in being amusing: his poem 'The Loves of the Triangles' was a parody of those of Erasmus Darwin, associate of Priestley and prophet of evolutionary theory later preached in prose by his grandson Charles. Frere adopted the character of 'Mr Higgins', whose beliefs were that 'Whatever is, is WRONG'; that 'if, as is demonstrable, we have risen from a level with the *Cabbages of the field* to our present comparatively intelligent and dignified state of existence, by the mere exertion of our own *energies*' we should, if Priestcraft and Kingcraft were abolished, 'enjoy unclouded perspicacity and perpetual vitality; *feed on* OXYGENE, and never *die*, but *by his own consent*'. Attempts to make radical changes in society were thus associated with evolutionary doctrines, with materialism, with Lavoisier's chemistry and with suicide.

'Higgins' hoped to teach geometry in didactic verse, thus strewing 'the Asses' Bridge [Euclid,I,5] with flowers', as Darwin had made attractive the sexual system of plant-classification:

> Thus, happy FRANCE! in thy regenerate land,
> Where TASTE with RAPINE saunters hand in hand;
> Where nursed in seats of innocence and bliss,
> REFORM greets TERROR with fraternal kiss;
> Where mild PHILOSOPHY first taught to scan
> The *wrongs* of PROVIDENCE, and *rights* of MAN;
> Where MEMORY broods o'er FREEDOM'S earlier scene,
> The *Lanthern* bright, and brighter *Guillotine*. [II, p. 204]

'Philosophy' meant science at this period; reform had been nurtured in England, but received in France the Judas-kiss of terror. We come closer to mathematics in these lines:

> But chief, thou NURSE of the DIDACTIC MUSE,
> Divine NONSENSIA, all thy Soul infuse;
> The charms of *Secants* and of *Tangents* tell,
> How Loves and Graces in an *Angle* dwell;
> How slow progressive *Points* protract the *Line*,
> As pendant spiders spin the filmy twine;
> How lengthen'd *Lines*, impetuous sweeping round,
> Spread the wide *Plane*, and mark its circling bound:
> How *Planes*, their substance with their motion grown,
> Form the huge *Cube*, the *Cylinder*, and *Cone*. [II, pp. 171ff.]

Here a footnote expands the lines about points progressing into lines with a sketch of Higgins' theory of evolution, in which 'the FILAMENT of *Fire* being produced in the Chaotic Mass, by an *Idiosyncracy*', vegetables would ultimately result; some of which 'by an inherent disposition to society and civilization, and by a stronger effort of *volition*, would become MEN.' Their tails would gradually rub off with sitting in their caves or huts. Such learned footnotes, sometimes dwarfing the text, were a feature of Darwin's didactic verse; and in Frere's muse 'Nonsensia' we see that genius for parody which later showed itself in Lewis Carroll and Edward Lear, whose productions are also an improvement on their originals.

Three months later the *Anti-Jacobin* [II, p.573ff.] produced a description of a new plant, *Directoria* belonging to the Pentandria group (the Directoire had five members) and the Polygynia order, to suit the morals of French revolutionaries. Written in Linnaean dog-Latin, it contains learned-looking footnotes directing the attention to the botanical writings of Rousseau, to the 'leafless and rotten stalk' called a tree of liberty that the French were wont to plant in countries they liberated, and suggests that the pistil of this plant is really a pistol. The seeds of *Directoria* are innumerable, mucus-generating, divisive, lancelike, and sorrow-making; the plant flourishes in Paris, will grow only if cultivated in Holland and Italy, and is absent from Kew Gardens; its properties are to produce vomiting and purging.

All this was a joke, but for those engaged in science it was a joke rather near the bone. John Robison of Edinburgh, where he was Professor of Natural Philosophy and editor of the lectures of the great chemist Joseph Black,[4] wrote *Proofs of a Conspiracy against all the Religions and Governments of Europe* in 1797. The conspiracy involved secret meetings of Freemasons, Illuminati, and Reading Societies; Jesuits and devotees of the mystic Jacob Behmen and of Swedenborg were also included. While a Masonic lodge in Britain was 'a pretext for passing an hour or two in a sort of decent conviviality', in Germany and in France lodges had become centres of immorality and sedition. The book makes rather dull reading after a while, because its hysterical tone is so continuous, and one immoral conspiracy described at length is pretty like another; but it gives some idea of the flavour of the 1790s, as that blissful dawn gave way to a bloody day and a protracted war. Bringing Louis XVIII to his own again was going to be a longer business than it had been with Charles II.

Our disposition to sympathize with the 'men of 1789' makes it difficult for us to appreciate the views of those in the 1790s who saw the revolution turn to tyranny, first by small groups and then by Bonaparte; and who saw the Revolution exported with bloodshed to the Netherlands and to Italy, who then had to pay the bill for their liberation. To Robison, some corruption seemed a fair price to pay for living in a free country.[5] It would be very wide of the mark to suppose that men of science in Britain were opposed to France and to Revolution because they were awed or bought by government. To be in favour of private property and to hate the French would be normal; but it did mean that science must be made the cement rather than the dissolvent of the bonds of society, and that the land of true freedom must be shown as apt for as good a crop of science as the abode of alleged liberty, equality and fraternity.

Materialism,[6] an evolutionary belief making man a mere animal,[7] and a dogmatism prepared to build systems in which conclusions were pushed far beyond the evidence:[8] these were the things to avoid. Their fruits were irresponsible social experiments. Indeed if with Burke one saw society as an organism rather than a mechanism, such experiments were a kind of social vivisection; and vivisection became taboo in Britain, though not in France. What was needed was a philosophy of science which went with stability and gradual reform rather than revolution, and which discouraged speculation; it was found in the home-grown writings of Francis Bacon.[9] His inductive philosophy, based on generalization of observations or experiments rather than on intuitive leaps or mathematical abstractions, became the norm in Britain – although, like most philosophies of science, it was little help if followed slavishly. Men of science presented their results as though it had been their method. This was particularly true in chemistry and geology, where 'sound philosophy' could exclude the notion that life was simply the outcome of inorganic processes, or that man had an evolutionary history.

Bacon also emphasized the usefulness of science. This was little more than a hope in Bacon's day, but by the late eighteenth century the Industrial Revolution was well under way. How much its early stages owed to science, and how much to a voluntary caste of intelligent Dissenters, excluded from full participation in official intellectual and social life, and following careers in industry and commerce, it is hard to say; perhaps the real remedy for Britain's industrial ills would be new Test Acts. But there could be little doubt that steam engines, and textile mills and gas lighting owed something to science; and Benjamin Franklin had invented the lightning conductor following his studies of electricity. Agricultural chemistry was just beginning; analyses useful to mankind, of mineral waters and of fertilizers, could be set against the French science which had culminated in the melting down of churchbells into cannon.

For Bacon too, in his *Advancement of Learning* it was 'an assured truth, and a conclusion of experience, that a little or superficial knowledge of philosophy may incline the mind of man to atheism, but a further proceeding therein doth bring the mind back again to religion'.[10] That the French were superficial was a comforting view for the English, though it hardly met the case in most of the sciences; but in Britain the tradition remained strong that the sciences were when properly understood a prop to religion. Bacon's idea was that there were two books that told us about God and His wisdom: the Bible and Nature. Study of the world might reveal no more than a First Cause or Supreme Governor, but a universe under the dominion of such a Being might be supposed to have some moral order built into it. Partly because the separation of science as a kind of profession came late to Britain, the link with religion remained strong right down to the middle of the nineteenth century; there is no need to suppose that the remarks about God in scientific books were not genuine. There were a few real atheists by the early nineteenth century, but to most people the events in France showed where atheism led, and men of science were careful to avoid it.

To pious Britons, it seemed that Providence had preserved the sceptred isle through all the hazards of war, indeed a world war, against the French; but the French engineer and physicist Sadi Carnot appealed to the other side of Baconian philosophy – technology, and in particular steam engines. For him, in his *Réflexions sur la puissance motrice du feu*, published in 1824, the little book in which the second law of thermodynamics was propounded,[11] wars were won by industrial strength rather than simply by the bravery and capacity of sailors and soldiers: and thus in a sense France, the leading nation in science, lost because she had been unable to turn this knowledge into power. Carnot's abstract and theoretical analysis of steam engines was of little use to contemporary engineers, and his views were lost to sight for twenty years. Meanwhile the unimportance of science in warfare was underlined by the French Institut's award of a prize for electrical discovery to Humphry Davy, with permission to go to Paris to collect it.

Davy, lecturing at the Royal Institution in London, became the apostle of applied science, urging the value of chemistry for tanning and agriculture, and the close association generally of science and industry.[12] It was unfortunate that his own connection with wartime industry, gunpowder manufacture, managed to lose money. More important than these lectures

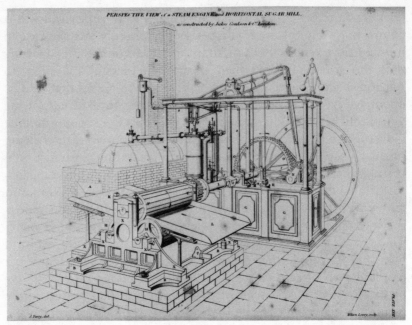

Figure 1   Steam engine and sugar mill from J. Fairey, The Steam
Engine, *1827 (reproduced by kind permission of The Bodleian
Library, Oxford).*

perhaps was his research leading to the isolation of potassium in
1807. Davy and his contemporaries inherited from Lavoisier the
idea that the elements in chemistry should be those bodies which
could not be further analysed. The list was thus provisional, for
some new method of analysis might break down compounds
previously resistant to decomposition. This notion superseded the
older one that the ultimate constituents of matter were atoms or
corpuscles, made of the same stuff, which yielded iron, gold,
sulphur and so on when arranged in various stable configurations.
The atoms had no specific chemical properties, possessing only
the 'primary qualities' of shape, size and mass; colours, tastes
and smells were the results of groups of atoms interacting with
sensitive organisms, and were called 'secondary qualities'.

Lavoisier reckoned that these atoms were hypothetical entities
which belonged in the realm of metaphysics rather than of
science, whereas his elements were empirical. He also believed
that the elements might carry properties. Oxygen was thus the

generator of acids, which is what its name means in Greek; anything containing oxygen, or containing enough of it, would be acidic, and without oxygen there could be no acidity. The question of the ultimate divisibility of matter was left to metaphysicians, or to physicists; the smallest particle of oxygen or of gold which a chemist might come across could be expected to have the properties of the larger quantities. Lavoisier's ideas, disseminated in his *Elements of Chemistry*, published in English translation in Edinburgh in 1790, were clear and persuasive and launched the science on an independent career; this made it in many ways the most exciting discipline of the early nineteenth century, where rapid advance in a new direction was the order of the day.

Davy, like many Englishmen, was interested in questions on the boundary of chemistry and physics: in electricity, heat and light, and in the nature of those attractive forces responsible for cohesion and for chemical affinity. He was unhappy with the new elements, and particularly with the idea that they brought properties to their compounds. His first juvenile essay was on heat and light, including some new observations in photosynthesis and the speculation that light as well as heat might enter into chemical combination. His next, and more disciplined, work was on the oxides of nitrogen, and appeared in 1800. He showed that oxygen and nitrogen in different proportions yielded air, which he like most contemporaries believed to be a loose compound: nitrous oxide, or laughing gas; and then three increasingly acidic, choking gases. The proportions and arrangements of components, rather than their nature, seemed to determine properties.

Davy's book made him famous not because of these analyses, but because of the descriptions it gave of the effects of breathing laughing gas. Davy was working from 1798 to 1801 at the Pneumatic Institution in Bristol, set up by the political radical Thomas Beddoes with Wedgwood money to see if Priestley's new gases might be useful in the treatment of respiratory diseases – especially consumption, the terrible killer of the young. Oxygen did not turn out to be as much help as had been hoped; and in trying carbon monoxide Davy nearly killed himself; but laughing gas, though some believed it would be instantly fatal, turned out to give all the pleasures of alcohol without leaving a hangover. Eminent Bristolians described their feelings for Davy to report. In

1801 Davy was invited to London, to begin his career at the Royal Institution, and at once displayed his talent for attracting and holding a large mixed audience.

At Bristol Davy had become interested in the electrical cell invented by Volta; he had recognized that a chemical reaction, rather than just contact of metals in a weakly acidic medium, was the cause of the current. If the combination of a metal with an acid generated electricity, then it was reasonable to suppose that, conversely, applying electricity to a compound might release the metal. Although Lavoisier had been unable to analyse potash or soda, which dissolved to give caustic alkalis, he had been unwilling to describe them as elements. He believed they were compounds, though he knew not of what elements. In 1806 Davy, by now at twenty-seven one of the leading men of science in Britain, was invited to deliver the Bakerian Lecture to the Royal Society and chose to work upon electricity in its relation to chemistry. In that year he showed that when water is decomposed by an electric current, the result is always oxygen and hydrogen in precisely the expected proportions. To demonstrate this he needed to take many precautions and to use apparatus of agate, gold and silver, because the glass of the time was not inert enough. In 1807 he was invited to lecture again, and this time, in a burst of activity, found that molten potash could be analysed with a current. The 'potagen', mysteriously reactive globules which floated on water and caught fire as they decomposed it, Davy decided were metallic; so he rechristened it 'potas*ium*', later potassium, in line with the conventions for naming elements.

Potash was an oxide, then; and Davy soon showed that soda was one also. Oxygen was therefore as much the generator of alkalis as of their opposites, the acids; and events soon seemed to show that oxygen was not in all acids. Hydrogen sulphide, the bad eggs gas, was a weak acid, although it contained no oxygen; and the even nastier-smelling hydrogen selenide was similar. The crucial case was the strong acid made from common salt; this was supposed to be composed of hydrogen, and of 'oxymuriatic acid', a greenish suffocating gas valuable as a bleach. There was no direct evidence that this gas contained oxygen if, as Davy showed, one excluded water from one's experiments; and he inferred that he was dealing here with an element rather than an oxide, rechristening the gas *chlorine*.

Instead of acidity being a sign of the presence of a particular substance, it became an effect of certain combinations, for hydrochloric acid, composed of hydrogen and chlorine, had nothing in common with sulphuric acid, which Davy and his contemporaries supposed to be composed only of sulphur and oxygen. It was not the matter involved which should interest chemists, but the forces which brought about fresh arrangements of that matter. For Davy, chemical affinity and electricity were manifestations of one force; that was why a chemical reaction could generate electricity, and an electric current could induce chemical change. Chemistry must be a dynamical science, and no longer a static one concerned simply with weights. Yet Davy's pupil Faraday succeeded, with his laws of electrolysis published in the 1830s in making dynamical chemistry also quantitative.

In a race with his great French rival Gay-Lussac, Davy showed that the mysterious substance iodine was analogous to chlorine. Gay-Lussac taught in a 'centre of excellence', the École Polytechnique, where he had been a star pupil, and he was a prominent academician at the Institut in Paris. Chemistry was still a French science, and Gay-Lussac produced a classic paper on iodine describing its various properties.[13] Davy's work could be described by an eminent contemporary as 'brilliant fragments'; he lacked the base, tradition and temperament required to systematize a science and to found a school, being more suited to revolutionary than to normal science. But his chemical work represented more than raids or forays into French science. Even though he could not work out all their implications, his discoveries led to fundamental changes in Lavoisier's paradigm. Gay-Lussac might have made this move, and had indeed toyed with the idea; but his loyalty to Lavoisier's chemistry, and in particular to the views of his 'father-in-science' C. L. Berthollet, who had done early work on chlorine, kept him back from this innovation. One might infer that in theoretical science the French were in the position of British manufacturers by the middle of the nineteenth century: having begun modern science in the one case and modern industry in the other, they found themselves in a generation or so overtaken by possibly more flexible rivals.

Where launching modern industry involved getting large numbers of people into factories, so launching modern science meant a change from the solitary genius to the trained member of a group or team. The French began the process whereby science changed from a hobby to a profession. At the École Polytechnique in the 1790s a student attended formal science courses delivered by men active in research; the best students might hope to join the staff there, or get a post elsewhere in which research might be carried on. The École Polytechnique was steadily made more military, and because the salaries attached to scientific posts were not large, the practice of *le cumul* – in which one man held several offices – grew up; nevertheless until the 1830s France remained the great example for British scientists of the country where science was properly supported. It is an irony that *le cumul* in science became prominent just when a similar practice in the Church of England was coming to seem a scandal and was being abolished; but defects in the system should not distract us from the transformation involved in formal courses. As medicine was being made more scientific, with the establishment of great hospitals in the later eighteenth century giving practical training to large numbers of surgeons, so science followed medicine in becoming something to be taught to classes. Many things that have to be learned cannot be taught, but science is not one of them.

Since the beginnings of the Renaissance medicine had been taught in universities to those who were to form the elite of the profession; and with law and theology it formed one of the three higher faculties in universities, to which students proceeded after taking the basic course. In the eighteenth century the great centres of medical education were Leyden, under Herman Boerhaave; Edinburgh, under the Monro dynasty; and Göttingen, where Albrecht Haller was Professor. The Leyden model involved both theoretical study of the range of medical sciences and also clinical work. Linnaeus at Uppsala had also been associated with Boerhaave, but his students tended to get their MD degrees for botany rather than for medicine in the strict sense. Anyone who had been through a course at a leading university would have got a sound grasp of medicine; but while quite large numbers could be taught in this way, most people about 1800 would never expect to consult a university-trained physician.[14]

It was the apothecaries and the surgeons who were the general practitioners of medicine, and they were trained much less formally. The basic route was apprenticeship. Only in the later years of the eighteenth century, with the birth of the clinic not only in France but also in Britain and Germany, did medical schools, generally private but often associated with a hospital, begin to train surgeons. Davy was apprenticed to an apothecary-surgeon in Penzance, then in the wild west of Cornwall; later he toyed with the idea of matriculating at Cambridge. An MD from unreformed Cambridge would have given opportunities for a career as a consultant at high fees, but he would have learned little medicine there. Some surgeons did go on to Edinburgh for serious work, or to Oxford or Cambridge, but most would never have aspired to it.

The learned professions by the end of the eighteenth century were controlling entry and standards of professional conduct, and were regulating fees, although curiously they all had two grades of more or less formality. In the law, the barristers were separate from and socially superior to the solicitors or attorneys; in the Church, there was a great gulf fixed between the higher clergy or those with good livings and the poor curates who did all the duties in many parishes; while in medicine the physicians were on a different level from the surgeons. From about 1800, on the other hand, the surgeon might have a better training than a physician who had not been to a first-rate university: he would have acted as dresser to a surgeon in a hospital, and would have also attended classes. John Keats and T. H. Huxley were notable products of this system.

There was nothing comparable in the sciences. Those interested in zoology or botany would have been well-advised to take a medical course, which had the further advantage that one could later support oneself by medical practice while doing science for fun; and since there was also a good deal of chemistry (for pharmacy) in the course, the same might apply to chemists. At Cambridge, mathematics was indeed the most prestigious course; but the highest honours went to those best drilled in the methods of Newton, ignoring the subsequent improvements in notation and methods made on the Continent. It was not until after 1815 that a group of reformers – notably Charles Babbage, John Herschel, and George Peacock – introduced French mathematics to Cambridge, making its course a good preparation for research in mathematics, astronomy or physics.

In fields such as geology there were professors who might give a course of lectures from time to time.[15] Lecturers should always remember that they are a part of the entertainment industry. Skilled performers, such as Beddoes at Oxford lecturing on chemistry, might draw very large audiences; but lectures which form no part of a formal course cannot go very deep. Most eminent men of science who had been to Oxford or Cambridge and found themselves interested in a science, had then in effect apprenticed themselves informally to the lecturer, helping with any research and going on what were essentially private field-trips, as the young Charles Darwin did. Only in the second half of the nineteenth century in England did it become possible to take a degree in science in the way which now appears to us so natural.

Davy, whose father was a wood-carver who died when Humphry was fifteen, could not have hoped to go to university; his formal medical apprenticeship was broken off when he went to Bristol to work with Beddoes on gases, and he went on to become a professional scientist, in the sense that he supported himself by his chemistry – in particular by lecturing at the Royal Institution in London, where he did his research. Faraday was a bookbinder's apprentice, who read some of the books he bound, and became increasingly fond of science.[16] In the days before about 1830 (when publishers started to sell books ready bound in cloth cases), books came in paper wrappers and binding them to customers' requirements was an important trade; science in contrast was a poor mistress able to support very few of her devotees. In 1812 a customer gave Faraday a ticket for Davy's lectures. Faraday made notes, wrote them up – unlike Davy he always wrote neatly – and delivered them to the great man, who duly found him a job making fair copies and washing bottles.

When in 1813 Davy went to France to get his prize, he took Faraday along as assistant cum servant. By 1818, when Davy was again away in Europe, Faraday in London was keeping him up to date and doing chores for him; Davy's letters now, instead of ending 'I am very sincerely yours, H. Davy', conclude with 'I am dear Mr Faraday your sincere friend & well wisher H. Davy'. Faraday seemed to have crossed the line from assistant to friend and colleague. A graduate student in the twentieth century would by that stage have submitted a thesis for a PhD, success involving a declaration of independence; but for Faraday, as for real sons of dominating fathers, escape was to be more difficult. In 1823 came a monumental row – these things were a feature of nineteenth-century science – when Faraday believed Davy had let him down over a priority dispute and then over his candidacy for the Royal Society, and the two never made it up.

Davy died in Geneva in 1829, and when in 1831 J. A. Paris published a *Life*, Faraday bound his copy in red leather and into it, as an act of filial piety, bound doodles, sketches, letters and original papers – remarking for posterity. 'It was my business to copy these papers. Sir H. Davy was in the habit of destroying the originals but on my begging to have them he allowed it and I have two volumes of such manuscripts in his own handwriting.' Davy, whose marriage to an heiress was childless and not happy, had another scientific son: his brother John, eleven years younger than he was and thus a contemporary of Faraday. He saw John through the medical course at Edinburgh, and into the Army, where he went on to become Inspector General of hospitals in the epoch of Florence Nightingale. He had been elected FRS in 1814, on leaving Edinburgh. He was the good son, attending Humphry on his deathbed and editing his *Collected Works*; Faraday, whose relationship was so much more problematic partly because he was such a very great scientist, found Davy a less good father but could not really escape from the relationship. Years of intimacy cannot be abrogated.

If Faraday had enjoyed fishing, which was Davy's greatest pleasure, they could have relaxed together and probably remained friends; in these quasi-family relationships small things can make the difference between rapture and rupture. Davy's 'sons' were bound to him by ties of kindred on the one hand and intellectual affinity on the other; and things were more difficult when the real son of an eminent scientist was less satisfactory to him than an adopted one. Thus C. L. Berthollet's son poisoned himself with carbon monoxide in 1810, recording his symptoms as long as he could; and Berthollet, who had himself been the protégé of P. F. Macquer, became father-in-science to Gay-Lussac. But there is a difference between the professor becoming close friend and patron of a favourite pupil and the researcher who makes his assistant his intellectual heir. The professor at the École Polytechnique and at institutions which followed it could have a large intellectual family, and his pupils could hope to move out and found their own intellectual families. This was an important feature in the Malthusian rate of growth of science during the nineteenth century.[17] An eminent professor would found a school, and fill the world with his descendants; and men took a pride in this kind of ancestry. Thus Benjamin Brodie, who was professor of chemistry at Oxford in the 1870s, was a pupil of Liebig, who had worked with Gay-Lussac and whose intellectual ancestry went back through Berthollet to Macquer.

Neither Davy nor Faraday founded schools in this sense. Their circumstances were not propitious, for without a system of scientific education the great man would have an only child or at most a small family, and in science as in natural selection success means leaving many descendants. Davy left Faraday to carry on with his interests on the frontier of chemistry and physics; but Faraday was childless both in his marriage and in his laboratory. Unless they fitted into a medical school or into the Cambridge mathematical network, men of science in Britain in the first part of the nineteenth century could hope to make friends and influence people in a general way, but not to build up a research school.[18] The need to appeal to outsiders reinforced the Baconian emphasis on the usefulness of science to the farmer, the manufacturer and the merchant. Faraday seems to have been fairly happy with the way things were, but many of his contemporaries looked across the Channel with envy.

Research schools are one thing, and general intellectual traditions another; but systems can reinforce orthodoxy. In France in the opening years of the nineteenth century the reigning tradition was the Newtonian one.[19] Berthollet and Laplace accepted Newton's picture of massy, indestructible particles of matter, so hard as never to wear away or break in pieces, and interacting through attractive and repulsive forces.[20] These particles or atoms were all made of the same stuff, and the chemical elements were just very stable arrangements of them: gold and iron differed in density because there were more or less particles in a cubic centimetre of each, and not because they were composed of distinct kinds of particles. Light was supposed to be a stream of minute particles, in line again with Newton's thinking; and to explain the polarization of light by reflection which he discovered in 1808, E. L. Malus (a young associate of Laplace and Berthollet in their Society of Arcueil) proposed that the particles of light were octahedral.

In France it was difficult to stand out against this orthodoxy; the very efficiency and centralization of things meant that a small group exercised a great deal of power and patronage in Paris, which was the centre of action. Newtonian heresies had more chance to develop in Britain, where patronage in the early years of the century was not yet much in the hands of men of science and there was no strong centre. In 1801 Thomas Young, who had studied at Cambridge, Edinburgh and Göttingen, after research on how the eye works, proposed a wave theory of light.[21] He put it before the Royal Society in a lecture in which he took texts in which Newton had written of frequencies, sides, and fits of easy and difficult refraction and reflection – picking out in fact places where Newton had recognized that without *ad hoc* hypotheses the theory of light particles will not work. In Young's work diffraction patterns fell into place as natural consequences of a wave theory; but because he followed the analogy of light and sound too far, some important facts about light could not be easily accounted for – such as its going in straight lines. Sound waves are longitudinal, the particles of air vibrating in the line of motion; a wave theory of light, however, works well only if the vibrations are supposed transverse, that is at right angles to the direction of motion. With this improvement the theory can account readily for polarization, when the vibrations are all in the same direction.

It was characteristic of Young, and a feature of science in Britain at the time, perhaps, that this full working out of the theory came later, from A. J. Fresnel and D. F. J. Arago in the France of the 1820s, when with the fall of Napoleon the grip of Laplace was loosened, and Berthollet had died. Young was a great beginner of things – the wave theory, the decipherment of the Rosetta Stone, a medical bibliography – which needed the more professional approach of the French to be brought to fruition. Heresy might be, like individual liberty, easier in Britain; but to develop into a new orthodoxy it needed the resources of France. The same kind of things can be said about John Dalton's atomic theory.[22] His suggestion that the atoms of all the elements were irreducibly different was again presented in a Newtonian manner, but it was clearly not what Newton had believed. The theory brought order into the understanding of chemical reactions, and meant that weights of substances entering into reactions could be calculated. That to do this brought about a certain theoretical incoherence did not worry the stubborn Mancunian who proposed it. Though Dalton's later work was not of much value to contemporaries or successors, it was when Gay-Lussac took up the idea of combining ratios, Gaudin distinguished between atoms and molecules, and Laurent gave it a hypothetico-deductive form, that Dalton's theory moved from being an interesting speculation into serious and testable science.

A half-way between the worlds of Gay-Lussac and of Davy was occupied by their contemporary J. J. Berzelius in Sweden. He taught a series of 'apprentices', who spent a few months or perhaps over a year with him, sharing his kitchen/laboratory and the services of his formidable housekeeper Anna. His students thus had the one-to-one relationship of father and son rather than of professor and student. Although they took no examination, the young men expected to leave after a time and take up posts elsewhere; several of them rose to great eminence while others, like J. F. W. Johnston, first Reader in Chemistry at the University of Durham, never became household names.

The wars between Britain and France also had more direct effects on some parts of science. Maritime rivalry had shown itself in the eighteenth century not only in warfare but also in voyages of discovery like those of James Cook and of Louis Bougainville – whose expedition recently became 'relevant' because one of his duties was to evacuate French colonists from the Falkland Islands under Spanish pressure.[23] These voyages brought accurate determinations of longitude and precision mapping – learned in Cook's case on the St Lawrence with

Wolfe – into the Pacific. Bougainville's voyage was followed up by that of La Pérouse, whose ships disappeared after setting sail from Botany Bay in 1788; although survey could be distinctly close to espionage, there was a convention that ships like these were exempt from molestation, and might call at hostile ports for supplies. The search for La Pérouse brought fresh ships into the Pacific; and then in 1800 Nicholas Baudin persuaded Napoleon's government to send an expedition to Australia under his command.[24] This, because of connections with the Museum of Natural History, was to be a highly scientific affair; but in the event Baudin had a series of rows with his scientists who always wanted him to stay longer at interesting places, and the voyage accomplished relatively little.

Citizen Baudin had two corvettes under his command, while the contemporary voyage from Britain under Matthew Flinders was sent on the cheap, in HMS *Investigator*, which turned out too rotten to complete the survey of Australia although it did complete a circumnavigation. Flinders, returning for a new ship, put in at Mauritius, where he was imprisoned by the French until the end of the war – conventions about survey ships being sometimes honoured in the breach. Accompanied by Robert Brown as botanist and by the botanical artist Ferdinand Bauer, Flinders had a highly successful scientific voyage.[25] When he was eventually released, Flinders wrote an account of it which contained a survey by Brown of the flora of Australia which is a landmark in the study of plant distribution. Superior French organization had in this case not led to better science; this was partly at least because of the lineage of Flinders, trained by Bligh, who was trained by Cook, and of Brown, the protégé of Joseph Banks, who had botanised at Botany Bay on Cook's first voyage. A strong tradition may be more important than liberal support.

Cook's naturalists on his second voyage had been J. R. Forster and his son George, who had written a book about it; their relations with the officers had been less than perfect, but they were competent naturalists and George was interested in ethnography. They had come from Germany to the lusher pastures of Hanoverian England; George went back to Germany, where he became a great celebrity, noted as a stylist. His example

fired the young Alexander von Humboldt, being trained as a
mining engineer, with determination to see the world, especially
the tropics. He went to Paris hoping to join a voyage to the South
Seas; but instead he found his way to Spain, then an 'ally' of
France, where he got leave in 1799 to go with A. J. A. Bonpland
to Central and South America. He returned to Paris with great
collections of material both scientific and economic, and having
climbed on Chimborazo higher than any mountaineer had gone
before; and in Paris under the Napoleonic regime he wrote up his
discoveries in thirty sumptuous volumes. These made him in effect
the founder of scientific geography. He appreciated that one
could indicate more than topography on maps, and also that
there were parallels between the vegetation of high mountains in
the tropics and of lower land in the temperate and even the polar
regions. His expedition, building upon traditions from Germany,
France and Britain, became the great model through the century.
His *Personal Narrative* of his travels fired the enthusiasm of
Charles Darwin.

Humboldt did not let the wars interfere with his work; though
as a Prussian, and therefore enemy alien, in Paris he was
subjected to some harassment, his friends and admirers in the
Institut supported him. Information about Latin America had
been treated as a Spanish state secret, and an indirect effect of
war had been Humboldt's expedition, which made it public
knowledge. A more direct effect was that the colonies of the
Netherlands, willy-nilly another ally of the French and even for a
period a French Département, were seized by the British. In Java,
Stamford Raffles, the founder of Singapore, was put in charge,
and wrote afterwards in 1817 his *History of Java*. Here again
information which had been a state secret, or rather a Company
one, was made public; and Raffles made enormous collections,
much of them unfortunately lost at sea. On his return to Britain,
in alliance with Davy, he founded the London Zoo in imitation
of the menagerie at the Jardin des Plantes in Paris.[26]

Rivalry with the French thus dominated British science during
the first thirty years of the century. Forays, some of them very
important, were duly made into French territory; but Paris
remained the centre of things. A feature of British science which
marked it as separate, or provincial even, was its emphasis on

natural theology. Science in France had emerged as a distinct activity or profession, justifying itself; in Britain it had not, and claims of utility meant both usefulness in industry and also support for belief in a Creator and Designer, and thus in a God-given morality. It is to this wrestling with God that we shall now turn.

3

Wrestling with God

The tradition that the study of nature tells one about its creator is a very old one. It goes back in literary form at least to Plato, and is therefore not specifically Christian. It became increasingly prominent in the scientific revolution, and Bacon made popular the idea that the Book of Nature could give the foundations of religion which would then be complemented by God's other book, the Bible, given us to reveal those truths which we could not learn from nature. If this were so, it became our duty as Christians to learn about nature, for natural religion was the cake on which revealed religion was the icing. For Bacon, science and religion were separate entities, and their practitioners should keep off each other's toes; but deep study of both would show how well they supported each other. Atheism was the fruit of shallow learning.

Bacon's example was followed by Robert Boyle in the second half of the seventeenth century. Boyle supported missionary enterprises and wrote works on philosophy and theology, as well as writing numerous scientific books. In his will of 1691 he left money to found lectures to demonstrate from nature the wisdom and goodness of God. In these a series of preachers expounded what came to be called astro-theology, where the heavens declare the glory of God, and physico-theology, where the evidence comes from animated nature. Newton's work on gravitation, because it indicated how simple and orderly the motions of planets and comets really were, was of much use to natural theologians, and Newton was happy to see his discoveries turned to this use. He himself was a deeply religious man, but not an orthodox Christian: early in his career he had convinced himself that the doctrine of the Trinity was a corruption of true Christianity, and he believed in a stern Father rather than a redeeming Son. These doctrines he kept from most contemporaries, for unitarianism was criminal down to the early nineteenth century. Natural religion was compatible with a series of religious beliefs going well beyond the boundaries of Christianity. This meant that while welcome to some divines, natural religion was regarded with grave suspicion by others. The danger as they saw it was that revelation might seem to be an optional extra, the natural religion being sufficient. Indeed there were authors of the early eighteenth century who urged just this, notably Matthew Tindal in his *Christianity as old as the Creation* of 1730 and William Wollaston in his *Religion of Nature Delineated*, for the 1725 edition of which Benjamin Franklin worked as compositor on his visit to London. If natural religion were a universal foundation, then one might well ask – as Fichte was to do in Germany in 1792 – how one would judge whether a revelation was genuine: if it confirmed natural religion, then it was unnecessary, and if it contradicted it, then it was false. Fichte fell back upon the need to remind mankind forcefully of the errors of their ways.

It has often been suggested that Puritanism had something to do with the rise of science.[1] Presumably it was not a factor in France or Italy, and by the later seventeenth century in England the chief effect of the Reformation on science seems to have been clerical marriage.[2] A supply of well-educated but poor sons of clergymen needing professional careers took up science; but by the time the Royal Society was founded their religious attitudes seem to have been liberal, or latitudinarian, rather than puritan. In the eighteenth century, high-churchmen – associated with the non-jurors who refused to take oaths to William and Mary, and with later attractive lost causes like the Stuarts – were highly suspicious of the mechanical tenor of much science, and its tendency to make God a mere First Cause in what was called Deism. They were staunch for the importance of doctrines and sacraments; this meant that they were more ready to see the world as a pattern of symbols than as an enormous clock.

John Hutchinson, a half-educated land agent, read a good deal of astronomy and geology and set himself to prove that the

account of the creation in Genesis was better than the best efforts
at explanation of Newton, Thomas Burnet and John Woodward.
His *Moses's Principia* was published in 1724–7; it and other
works of his attracted some disciples, especially because of his
interests in symbols and prophecies. Questions of Biblical
interpretation had indeed agitated Newton too, who had been
particularly interested in decoding the *Book of Daniel* in the
same sober and literal fashion in which he had tried to make
sense of alchemical recipes. The most important of Hutchinson's
followers was William Jones, perpetual curate of Nayland in
Suffolk. Hutchinson had been struck with the parallels between
the union of fire, light and air, and the Trinity; Jones too wrote
on the doctrine of the Trinity in 1756. He was a leading high-
churchman of the day, later praised by Newman; but he
left advocates for his theology and for his science, the
most noteworthy being William Kirby, the leading entomologist
of the 1820s and author of work on natural theology; and
William Swainson, a younger speculative zoologist and a superb
illustrator.[3]

Those in this tradition saw analogies everywhere in nature.
Everything was bound together, and sermons in stones or
arguments for the resurrection in insect metamorphoses were a
source of delight and consolation. A sacramental religion easily
led to a resolve to see ordinary things as a glass through which
the heavens might be spied. Those who thought this way saw
inspired texts of the Bible as true, but not in a simple-minded
way. In the long tradition of rational, rather than rationalistic,
interpretation they saw truth on different levels in different
places – poetic as well as literal truth – and faith more as
believing in students than believing six impossible things before
breakfast. As compared to evangelicals, high-churchmen ran the
risk of being high and dry, relying entirely upon the head rather
than upon the heart; but they were in no risk of Bibliolatry, and
thinking is no obstacle to the pursuit of science.[4]

Hutchinson and Jones saw analogies everywhere, so much so
that definite falsifiable explanations seem hardly to have
interested them. Their more-scientific heirs were to find this a
problem. But a prominent churchman in a more central tradition,
Bishop Joseph Butler, in his *Analogy of Religion* of 1736,

emphasized the analogies of nature for religion in a more general way. An effect of his book was to make many working in science aware of the analogical nature of much reasoning used in scientific discovery. Butler's book is in two parts, one being devoted to natural and the other to revealed religion. In his famous Advertisement at the beginning of his treatise, he wrote: 'It is come, I know not how, to be taken for granted, by many persons, that Christianity is not so much as a subject of inquiry; but that it is, now at length, discovered to be fictitious.' This was an exaggeration, but his book turned the tide and he was generally respected as a sound authority. In England intellectuals in the later eighteenth century were not dismissive of religion as they were in France.

Butler's discussion of probability is particularly interesting, for nineteenth-century science was to become increasingly a matter of probable rather than demonstrative evidence. In the same way, his remark that the whole argument is cumulative applies to work such as Darwin's where no individual case may be convincing but the weight of them all is hard to resist. The convergence of lines of argument in science is very persuasive, and in the nineteenth century convinced sceptics that matter was composed of atoms, for example. For Butler, 'probability is the very guide to life'; and this was necessarily so because we were beings of limited capacities. Only to God could there be certainties, because only He knew all the circumstances of every case. For Butler then, science was provisional; it should avoid extravagant hypotheses, working by analogies, and one of its major functions and pleasures should be the uncovering of God's handiwork.

In Scotland the rational theology was Calvinist rather than high-church; and there, in David Hume, came the great critic of natural theology, particularly in the posthumously published *Dialogues concerning Natural Religion* of 1779.[5] The book achieved a certain notoriety, but it seems that in the eighteenth century Hume was not taken seriously as a philosopher. To us his remarks that one cannot really compare the world to a clock and infer that both had a designer; that animal suffering (which cannot, like human, be redemptive) is hard to reconcile with benevolent government; and that nature can easily be seen as red in tooth and claw, all seem persuasive, and the spokesman for natural religion seems bland. Hume demonstrated that one can as easily see in nature contrivances which seem designed to cause pain as to give pleasure: the ichneumon fly and the liver fluke are nature's missiles, and there is no easy way to prove a benevolent designer, or to find a morality in nature. But Hume did not have very much effect in Britain until his philosophy, having woken Kant from his dogmatic slumbers, came back in German dress in the nineteenth century; and it then took the form of a theology in which 'evidences' from nature were played down.

Because apologists for natural religion realized that there was pain and suffering in the world, their argument depended on showing that on balance there was more pleasure than might have been expected – which is not much consolation for the widow, the orphan or the refugee. This meant that they were concerned with a calculus of pains and pleasures, which is found in Wollaston; and that their arguments went readily with a utilitarian ethic. The term 'the greatest good of the greatest number' was first used by Joseph Priestley and is generally associated with Jeremy Bentham; but at the opening of the nineteenth century a leading exponent of utilitarianism was William Paley, now chiefly remembered for his *Natural Theology* of 1802.

Paley's famous beginning summarizes the whole argument: that if we should happen to come across a watch, and see how all the parts were made of suitable materials for the function they had to perform and how everything worked together, we should not take seriously the suggestion that its atoms happened by chance to have come together in this way. This is a version of the argument that a monkey at a typewriter could not write *Macbeth*. Paley then seeks to demonstrate that the whole world is like a watch, and particularly that organisms are. He believed that God had added pleasure to things which might have been merely neutral, like eating; and watching the shrimps playing in the then-unpolluted waters of Sunderland (where he was a parson) he believed that they too were enjoying themselves. There is much in Paley of which one can make fun in the twentieth century. He was not a man of science, and was summarizing the work of predecessors; there is a certain complacency in his tone which exasperated some contemporaries; but he made sense of a great deal of scientific information, in a kind of synthesis. His book was compulsory reading for undergraduates at Cambridge in the first half of the nineteenth century, and also at the new university of Durham from 1832. It gave those destined for the Church, or attending an intellectual finishing school, a scheme for fitting scientific knowledge into their world view. In so far as there were not 'two cultures' in nineteenth-century Britain, Paley can be given some credit.

Natural theology was not limited to high or broad churchmen in England, but also went with Calvinism. In America, Cotton Mather published his *Christian Philosopher* in 1721; and in 1817 Thomas Chalmers of Glasgow delivered and published a *Series of Discourses* on Christianity and modern astronomy. On the effects of missionary preaching to the Greenlanders he had written in 1813: 'The demonstrations of natural religion fell fruitless and unintelligible upon their ears; but they felt the burdens of sin and of death, and pleasant to their souls was the preacher's voice, when it told that unto them "a Saviour was born".' For a more sophisticated urban audience, some natural theology would be in order, even from an evangelical Calvinist; and the *Discourses* were to meet the threat of infidelity from astronomy rather than to provide a natural religion.

In speaking of Newton, he referred to his 'march of intellect', a phrase which became a cant term for the diffusion of rather low-grade 'useful knowledge' in the 1820s. But it was to misconceptions of Newton that he addressed himself. It was not the lustre of superiority which was his characteristic, but peaceful, unambitious modesty. Science went wrong when it was dogmatic, when it forgot its provisional character; when 'philosophers . . . have winged their way into forbidden regions – and they have crossed that circle by which the field of observation is inclosed – and there have they debated and dogmatised with all the pride of a most intolerant assurance'. Vague analogies had led to the production of what were no more than amusing philosophical romances, or what we would call science fiction. Baconian philosophy thus fitted very well with natural theology, in its emphasis upon cautious generalization from observation, whether of an evangelical or an episcopal kind: it thus helped to reinforce that empiricist rhetoric which is so characteristic of men of science in the early nineteenth century.

Chalmers' chief target was no doubt Laplace, the man who 'had no need of that hypothesis' that God was needed to keep control of the planets in case they wandered from their courses, and who in his work on statistics seemed to assume that men were not free moral agents but subject to chance. In Britain, Laplace had few followers; partly because, in this provincial region, there were few who could follow his mathematics – his great book was translated into English not by a Cambridge mathematician but by an American sea-captain, Nathaniel Bowditch – and partly because most natural philosophers were more or less religious. Many of the leading men were in fact clergymen, because most dons at Oxford and Cambridge were: ordination was a common step in an academic career. There was certainly some anti-clericalism, especially among the younger generation like the geologist Charles Lyell, but very little anti-religious sentiment down to the middle of the century.

Methodism had transformed the Nonconformist churches in England in the latter years of the eighteenth century, and in the opening years of the nineteenth it worked in the Church of England too in the form of the Evangelical Revival. This led to a greater stress upon the Bible; and at the same time the rise of geology meant that it seemed likely that the account of creation in Genesis might be confirmed, or by speculative and dogmatic geologists denied.[6] Geology was not the only science offering such a test: the astronomy of Copernicus, Galileo and Newton was not readily compatible with the literal text of some Biblical passages, and linguistics made it less easy to trace the Tahitians, the Basques and the Algonquins all back to the Tower of Babel. On the other hand, most of the more eminent evangelicals were not what might be called today fundamentalists. Among those who wrote on geology were Hugh Miller, the friend and associate of Chalmers in the Disruption of the Church of Scotland in 1843; John Bird Sumner, future Archbishop of Canterbury; and Adam Sedgwick, Professor at Cambridge and Canon of Lincoln. All of them expected the strata and fossils to confirm in a general way what was in Genesis, but certainly did not start geologizing from the Bible.

In Hume's freethinking Edinburgh circle, James Hutton had in 1788 published his essay *The Theory of the Earth*, which ended with the famous words: 'The result, therefore, of our present enquiry is, that we find no vestige of a beginning, – no prospect of an end.' This essay was later expanded into an unreadable book. Hutton's obituarist contrasted him with Black, who 'dreaded nothing so much as error', while Hutton 'dreaded nothing so much as ignorance'; certainly he set a cat among the pigeons with his speculative essay. And yet his concluding phrase was meant not as a challenge to Christian doctrine of the creation and the end of the world, but as an avoidance of explaining first origins of things. Geology was to be based on reasoning from distant events, on going back from effects to causes, and could not deal with things outside the steady cycle of decay and renovation. Through Hutton's writings comes a steady note of praise for the First Cause; but this is some way from Christianity, and Hutton's refusal to look at first origins meant separating geology from religion. The geologist drew his evidence from the quarry and the outcrop, and paid no attention to the Bible.

Hutton's endless past time was not quantifiable; one might expect that on past continents there had been animals and plants, and indeed there were 'monuments' which proved it. Geologists liked to use terms like Monuments and Medals to describe fossils, thereby linking their science to archaeology, which in the latter part of the nineteenth century became in turn indebted to geology for its scientific basis. But Hutton saw no reason to doubt the recent creation of man, as set out in Genesis, for there were no human fossils known. Fossil hunting in Hutton's day was still unsystematic, like treasure hunting; and it was only about 1800 that, once again from France, the new science of palaeontology entered geology and transformed it.[7] One can see this in the *Organic Remains of a Former World* by James Parkinson (describer of Parkinson's Disease) published in 1804–11. The frontispiece to the first volume shows Noah's Ark, with some ammonites and trilobites which had missed the boat lying dead and ready to be fossilized; and the text is concerned with justifying the notion that fossils are the remains of creatures which lived a long time ago, and that some kinds of creatures have really become extinct.

These were somewhat alarming notions in the last years of the eighteenth century, for they implied a very long history for the Earth. If the point of the Creation were to provide a sphere of probation for mankind, where people could prepare themselves for eternity, then one might ask why they appeared so late upon the scene. An answer worked out in the nineteenth century was, so that they could have coal to keep warm and to fuel steam-engines. But one might expect that if at one time it had been good to have mastodons in Ohio, then it should always be good, and a benevolent Deity should not have allowed them to become extinct. Extinction also threatened the idea of a great Chain of Being leading up from crystals through amoebas and corals to animals and on to man; if links in the chain were to be missing, then the world would be impoverished and the order of things no longer clear.

In Parkinson's later volumes, the tone changes, for he became aware of the work of Cuvier, whose work appeared in various papers and finally in the *Récherches sur les ossements fossiles* of 1812.[8] As Paris was rebuilt under Napoleon, limestone was quarried in Montmartre and was found to be full of fossils, notably of mammals. Cuvier found that although it is rare indeed to discover a complete skeleton, he could with experience decide where any bone should go. Not only did he reconstruct extinct rhinoceroses and bears, he also found a series of distinct faunas. The Paris basin had had successions of creatures living in it; it seemed to Cuvier that the lines separating these different populations were clear and sharp, and must indicate catastrophes which wiped out one fauna, making room for another. The discovery of the first dinosaurs, by Buckland and by Mantell in England, added to the list of past faunas.[9] The finding of mammoths frozen in the Siberian ice, and indeed quick-frozen, for they were still edible, implied that the catastrophes must have been sudden. Noah's flood seemed one among a series of dramatic episodes, the rest of which had happened before the creation of man.

It could hardly be claimed that the existence of dinosaurs, any more than that of stars invisible to the naked eye, really threatened the authority of the Bible, which was not required to be more than sufficient for salvation. The crucial question

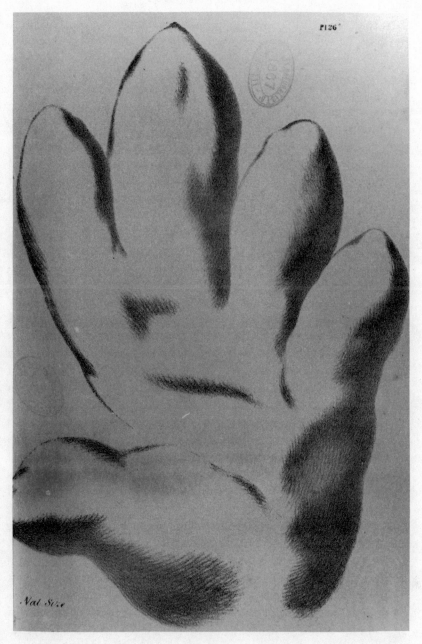

Figure 2 *A dinosaur footprint from William Buckland's*
Bridgewater Treatise, *1837.*

therefore was the recent appearance of man, at approximately the date which could be computed by adding up the ages of the patriarchs: the famous 4004 BC. Cuvier devoted some space at the beginning of his book to criticizing the chronologies of the Hindus and others which appeared to go back beyond this date. Actual harmonizing of the description of events in Genesis and the findings of geologists remained complicated, partly because the accounts in chapters 1 and 2 are not consistent, and because of all sorts of problems about the meaning of the days of Creation. Either they might be supposed a thousand ages long, or else it could be urged that long undescribed periods intervened between the days of Creation and the appearance of mankind.

In 1807 the Geological Society had been set up in London, devoted to empirical investigation and eschewing speculation, whether of a Huttonian or a scriptural kind. Most societies devoted to a single science in the nineteenth century also adopted a Baconian stance of this kind, aiming for scientific respectability by postponing hypothetical reasoning; and the early papers published by the Geological Society were very descriptive.

By the 1820s, William Buckland of Oxford seemed suddenly in a position to cast light upon the Flood with the discovery of bones of hyenas and other creatures in the Kirkdale Cave in Yorkshire.[10] His book describing and interpreting the finds, *Reliquiae Diluvianae*, 1823, was a great success in its day; and Buckland succeeded in fitting evidence from other countries into the pattern, indicating what seemed to him a great flood in the relatively recent past. Much of his evidence, from erratic boulders and from U-shaped valleys, was later to be interpreted in terms of ice action; but the most interesting parts of Buckland's book are those where he compared the past hyenas with modern ones. He went to Wombwell's travelling menagerie when it came to Oxford – there was as yet no London Zoo – in order to watch a modern hyena crunch up bones, and to compare what he left with the Kirkdale bones. In a disagreeable experiment he also compared 'many small balls of the solid calcareous excrement of an animal' found at Kirkdale with the faeces of a living hyena, which turned out to be precisely similar. At Kirkdale, therefore, there were not just stray bones all washed together: the cave had been a hyenas' den, where they had lived and gnawed the bones of their prey. Antediluvian Yorkshire had had a fauna more like that of Africa than of modern England.

In this work, Buckland like Cuvier was doing his best to interpret the past using knowledge of what happens in the present. This had been an axiom of Hutton's, and Buckland's pupil Charles Lyell made it the explicit foundation of his *Principles of Geology*, which began to come out in 1830 and was taken by the young Charles Darwin on the *Beagle*. Lyell used this idea in order to challenge the notion that the history of the Earth was a series of revolutions with quiet periods interspersed – rather like the history of modern France. He thought his predecessors, in a famous phrase, had been prodigal of violence because they had been parsimonious of time: that is, that what seemed like the effects of catastrophes could be produced by ordinary processes operating steadily for millions of years.

This idea was alarming to some, but it did not stop Lyell being appointed Professor of Geology at the newly founded King's College, London, which was meant to be the Church's answer to the secular University College. (The two eventually formed a federal university.) Nevertheless, about 1830 the anxieties Chalmers had felt about infidel astronomers were being transferred to geologists. Lyell was all right because he stressed the recent appearance of mankind, and because he specifically attacked the evolutionary notions of Lamarck; but in his kind of geology there was no longer room for one mighty Flood showing its effects all over the world, but only for series of local floods of the kind we are familiar with.

Just at this period, in 1829, the eighth Earl of Bridgewater – Francis Henry Egerton, heir to a fortune from canals and mines – died, leaving £8,000 to be given to one or more authors, chosen by the President of the Royal Society, who would write on 'the Power, Wisdom and Goodness of God, as manifested in the Creation'. Bridgewater was himself a clergyman, a Canon of Durham (where his father had been Bishop), who lived in Paris surrounded by cats and dogs, some of which fed at his table and rode in his carriage. The President, Davies Gilbert, consulted the Archbishop of Canterbury and the Bishop of London before inviting eight authors to write treatises, which duly appeared during the 1830s.[11] This decade of political excitement and change thus saw the flowering of the conservative-seeming programme of natural theology. Paley's work had become out of date, despite revisions by later editors, some of whom even appended specimen questions at the back to help students mugging up; but none of the new authors wrote as well as Paley.

It is tempting to see the Bridgewater authors as a group having a common party line; to some extent they did, but their theological views and their posts were very different, and they should not be seen as a kind of party. The first treatise was by Thomas Chalmers, and it was seen as a sign of liberalism that a Presbyterian should have been thus chosen: for this was the period of Roman Catholic Emancipation, and of fear that with the Reform Bill the Church of England would be in real danger of disestablishment and confiscation of revenues. John Keble's Assize Sermon on national apostasy was preached in 1833, and 'The Church is in danger' was a cry to move any loyal Anglican. In 1832, the University of Oxford had even conferred honorary degrees on four eminent scientists, Robert Brown, Dalton, Faraday and David Brewster, none of whom were members of the Church of England. Buckland and John Kidd, Professor of Medicine at Oxford, were Bridgewater authors and might be seen behind this liberalism, striking at a time when all undergraduates had to be Church members.

The most interesting treatise from the point of view of originality was that of William Whewell, of Trinity College Cambridge.[12] Whewell was one of the great pundits of early Victorian England, expert in mineralogy, theory of the tides, German Gothic architecture, and moral philosophy. He was a great coiner of scientific terms and also of the word 'scientist'; and he was one of the first to take the point made by Hume and by Kant that the study of nature cannot lead to a proof of the existence of God. He adopted a different approach, in which the believer came to science with an idea of God's wisdom and benevolence which could be tested and refined by contact with the facts. He went on in his writings on the history and philosophy of science to generalize this, in opposition to the prevailing Baconian outlook. He recognized that there is a leap to be made in going from instances to a law or theory; and suggested instead that what is needed is to begin with an appropriate idea of the basis of a science, and to refine it by experimental tests. For him, science therefore had to have a metaphysical basis, and could not rely simply on observations and generalizations from them; and in his mind this was closely connected with his natural theology.

Oxford and Cambridge were clerical institutions, still in the 1830s almost seminaries; and another of the authors was a clergyman, William Kirby. He was a high-churchman, an admirer of Jones of Nayland, whose attempts at decoding the Scriptures he continued in his Bridgewater Treatise, on the 'history, habits, and instincts of animals'; and was thus very different in outlook from the evangelical Presbyterian Chalmers or the liberal Tory Whewell, being orthodox and anti-Calvinist. While reasoning about the Deluge, the relationship between the Cherubim of the Old Testament and the Forces and Powers of nineteenth-century natural philosophy, or the connections between souls and matter, Kirby seems remote from us; but like other orthodox divines, he was insistent that whatever wanderings different groups might have gone in for since leaving the Tower of Babel, all mankind was of one blood. The opposite belief, that different races were distinct species, could be used to vindicate black slavery; so in the nineteenth century one's attitude to the doctrine that all mankind was made in the image of God mattered.

Kidd was both a doctor and an Oxford professor; and the last three Bridgewater authors were all medical men. Sir Charles Bell, a Scottish Episcopalian, was a very distinguished physiologist, who had demonstrated the distinction between the endings of sensory and motor nerves. This had involved him in a race with François Magendie in Paris, who had not hesitated to use vivisection; and the whole affair resulted in an international row in which true Britons supported their man against the 'murderous Magendie'. Bell wrote on the human hand, for the Earl of Bridgewater had been persuaded that the hand offered particularly good evidence of benignity; but he used the book to get across much physiology and comparative anatomy. P. M. Roget (whose treatise was on physiology) is now remembered for his *Thesaurus*, without which no crossword puzzles should be attempted; this was based on a system he had devised for organizing knowledge. He was for many years Secretary of the Royal Society, and presided over the launching of its less-formal journal, the *Proceedings*. William Prout, who discovered hydrochloric acid in the stomach and thus gave a rationale for antacid medicines like milk of magnesia, had a rag-bag of 'chemistry, meteorology, and the function of digestion' to cover.[13] He had some interesting things to say on atomic theory, having by the 1830s modified his famous hypothesis that all the chemical elements are composed of hydrogen atoms in different arrangements; but he still refused to believe that there could be over forty different basic particles.

Whewell's Bridgewater Treatise provoked the Cambridge mathematician Charles Babbage to write a riposte, not because of his general approach but because of remarks about the low value of purely deductive reasoning. Whewell, writing about astronomy, was anxious to show how poor Laplace's judgement of religion had been, and blamed this on his excessive use of mathematical reasoning: Whewell hoped for a middle way in science, neither Baconian introduction nor Cartesian deduction. Babbage believed that the mathematical sciences had not had proper representation in the Bridgewater Treatises (as they had not in the scientific establishment), and his unauthorized *Ninth Bridgewater Treatise* was to get the balance right; it was also a contribution to his campaign to prove that science was in decline because mathematics was not sufficiently supported. The striking thing about his book is that he imagines the world as a great computer whose programme we are trying to guess, so that what seem to us miracles – arbitrary exercises of Divine power – may be better understood as programmed in. Thus one might save the ideas both of divine government and of Laplacean determinism. Babbage spent much of his life in the attempt to construct enormous clockwork computers, until the government in an economy drive cut off his funds.

Buckland in his Bridgewater Treatise gave up the idea of a universal flood, and his book is not at all an exercise in fitting Genesis to palaeontology; he remarks that the phrase 'in the beginning' covers the whole of geological history down to the appearance of man. He was most concerned with showing how well-designed extinct animals were. He believed that fossils called coprolites were the faeces of extinct creatures, from which one could reconstruct how their bowels moved; in general he was anxious to show that they had died out through no fault of their own. Even the disturbing agencies which had destroyed extinct forms of life showed design, which was found in undeviating unity throughout creation.

Buckland's books were sometimes laughed at by his colleagues: his *Reliquiae Diluvianae* was described as 'Idola Specus' (the Idol of the Cave, against which Bacon had written), and his Bridgewater Treatise as a 'Bridge over the Water'. He nevertheless had a high reputation generally, and carried liberal-minded churchmen with him in believing that Christianity and science were in harmony: he himself was duly made Dean of Westminster. He was never pompous, and was famous for his serious clowning at scientific meetings, when he might get down on all fours to imitate the movements of some animal whose fossil footprints he had found. Outside the circles of science, there were 'Scriptural geologists' who sought to make an exact fit between the Bible and the quarry; the most eminent was Dean Cockburn of York. It was important for those with a place in both the scientific and the church 'establishment' to distance themselves from literalists on the one hand, and from materialists on the other. On the whole in the 1830s and 1840s this could be done without great difficulty – especially if one had a Baconian vision of science, in which both opposition parties had strayed from the paths of sound induction, following delusive preconceptions.

Buckland was happy to quote a long note from his Oxford friend E. B. Pusey on the interpretation of the Hebrew word translated 'create'.[14] Pusey was to become after 1835 one of the leaders of the Oxford Movement; and later in the century high-church clergy who emphasized ritual in worship, and regarded themselves as a Priesthood set apart from laymen in a way parsons of the early nineteenth century had not, were called Puseyites. Although Pusey became increasingly conservative, in the early 1830s he was a pioneer of the scholarly study of the Bible (and later of the Church Fathers), bringing to England a badly needed dose of German scholarship; this in liberal hands by the middle of the century seemed a threat to old-time religion, but not in the 1830s. There was some alarm about science, but there was no general feeling that it mounted a great challenge to the Church beloved of high-churchmen, or to the Scriptures reverenced by evangelicals.

Pusey had come to be a colleague of Buckland's at Christ Church, Oxford, from Oriel College; and in the 1820s under Edward Copleston, Oriel had become the leading college in the university, stressing learning among Fellows and undergraduates. Copleston had come to prominence when in 1810 he published his *Reply* to the 'calumnies' of the *Edinburgh Review*, where Oxford had been unfavourably compared to the universities of Scotland and Germany. Articles by Richard Payne Knight the aesthete, John Playfair the mathematician and Sydney Smith the wit had assailed Oxford as a wealthy stick-in-the-mud; and Copleston sought to demonstrate that their charges were out of date or ignorant, and that in criticizing a recent Oxford edition of Strabo, Knight showed he knew little Latin and less Greek. Copleston argued particularly for a teaching university, rather than one where research was paramount, based on tutorial teaching rather than professorial lectures, and for teaching a common culture of established classics rather than encouraging young and untrained minds to speculate. 'Never let us believe,' he wrote, 'that the improvement of chemical arts, however much it may tend to the augmentation of national riches, can supersede the use of that intellectual laboratory, where the sages of Greece explored the hidden elements of which man consists, and faithfully recorded all their discoveries.'

The debate about usefulness, and the place of science in a liberal education, was to be carried on throughout the century, not merely in Britain but also in Germany. In this particular controversy, coming at the heyday of Edinburgh's prestige and being one round in the England/Scotland war of words, both sides seem to have felt that they had won: one's judgement of the result depended on which side of the border one inhabited. But two visions of universities were presented, and which one was followed was to make a major difference to the state of science. The word 'science' in Copleston, in the *Bampton Lectures* (1815) of William Van Mildert of Christ Church, and in their contemporaries was used of any organized body of knowledge; it was opposed occasionally to 'literature' but generally to rule of thumb, or 'empiricism' – which was then associated with empirics, or quacks, rather than with Baconian philosophy. The sciences, in our narrow sense, were not separated from other disciplines; and the expectation was that after a liberal education, one might specialise – in Law, in Theology, in Medicine or in Natural History, with different career prospects in each case. The specialized study might be done at a university, but might well be a kind of apprenticeship: Faraday learning from Davy was not so different from a young curate learning from an experienced vicar, or a young barrister doing his pupillage.

The most eminent beneficiary of the Oriel system of granting fellowships following competitive examinations was J. H. Newman, the future Cardinal. The Oriel men were known as the Noetics because they were believed to be interested in the intellect alone; which does not seem true of Copleston, and is certainly untrue of Newman and his friend John Keble, whose devotional poems *The Christian Year* (1827) were exceedingly popular. In his Oxford University Sermons, which electrified undergraduates in the 1830s and were preached while he was still an Anglican, and then much later in his *Grammar of Assent* (1870), Newman struggled with highly-sophisticated notions of knowledge, proof and demonstration.[15] He stressed the importance of personal knowledge; the need to commit oneself to something; the low value of 'notional assent' to propositions which we do not care about either way; and how faith and reason are not in competition but should complement each other. Newman's book contains a good deal of religious apologetic, which has sometimes diverted attention from its philosophical underpinning; here, as in his famous *Essay on the development of Christian doctrine* (1845), Newman was showing himself to be an important philosopher of religion, gaining something from a very wide view of science. In our narrow sense of science, belief is an essential component; and Newman's insights are an excellent corrective to the view that any body of real knowledge is a mere collection of authenticated facts. The gritty idea that any statement must be true or false would be far from the mind of Newman; but this generally admirable quality, and his respect for doubt, made him seem slippery to the more stolid.

Newman had little time for natural theology, as a dry business which could lead only to notional assent; whereas true religion was a matter of real assent, or commitment. Here the poet and liberal theologian S. T. Coleridge would have agreed with him, although Coleridge was excited by some parts of science, such as Davy's demonstration that forces rather than matter were crucial in chemical explanation.[16] Coleridge saw materialism, which in his youth he had embraced, as a great bugbear; and he was delighted that what seemed to him the best modern science had removed all plausibility from these gloomy doctrines.

Similarly, in the *Discourse on the Studies of the University* (1833) which began as a sermon by the evangelical geologist Adam Sedgwick preached in 1832 at Trinity College, Cambridge, we find faith presented as a rule to live by, and not as assent to propositions. Indeed, for churchmen of whatever party, faith was akin to belief in Britain rather than to belief in Father Christmas. Sedgwick assailed mere empiricism and also utilitarianism; and he vindicated natural theology, giving praise to Whewell's *Bridgewater Treatise*. His book, defending a liberal education in which mathematics was far more prominent than in Oxford, was popular; and when in 1844 an anonymous work, *Vestiges of the Natural History of Creation*, came out he added an appendix with the intention of refuting it. Such was his outrage that the appendix dwarfed the original text, and its remarks about the possible corruption of innocent Victorian womanhood are now sure to provoke mirth; but *Vestiges* was one of the first really influential and important works in which in the name of science theologians were thrown on the defensive, by what seemed a persuasive picture of materialism and determinism.[17]

The author was Robert Chambers, a Scot and publisher especially of educational works; but whereas the real authorship of most anonymous Victorian writings, like the articles in the *Reviews*, were open secrets, Chambers' was disclosed only after his death, forty years later. *Vestiges* is a curious work, presenting an evolutionary theory beginning with a 'fire mist' out of which the Sun and planets had coalesced. A First Cause presided over the development according to law of living forms, and their progress from single-celled creatures up to the best so far, mankind. Development happened when embryos stayed longer on the main line, and we might expect that in the fullness of time some human mothers will give birth to a prodigious new species, to whom we shall seem like apes. We think we are choosing freely, but recent statistical work (notably on the murder-rate in Belgium) showed Chambers how predictable mankind was in the mass:[18] we are really subject to iron laws like everything else.

Vestiges was denounced by almost everybody, especially by expert biologists and astronomers who were appalled by its blunders in details and by its desertion of Baconian principles in favour of speculation. The clergy hated it for its apparent denial of human responsibility, and feared that those who were told they were animals would ignore the moral code. All this meant that it sold very well, reaching its twelfth edition in 1884. What is striking is that some, like T. H. Huxley, who welcomed Darwin's *Origin of Species* in 1859, had assailed *Vestiges*. It was not the materialism which appalled him, but the amateurism of the book; and it is always a problem for anybody with some very new idea to convince sceptical colleagues that he is on to something important. Science is a deeply conservative business, and most bright ideas turn out wrong; but the problem of convincing sceptics in a Baconian age was acute. Some saw hope in Germany, where Kant's 'Copernican Revolution' in philosophy had set subsequent thinkers on a new track and where philosophy was the normal introduction to science. Others saw Germany as the home of pedantry and misleading metaphysics; but at least all agreed that the flippancy and positivism of the French was not to be found there. In this serious-minded but still bucolic country, there were arguments about the nature and drift of science and of education which were of great importance for the rest of the century.

4

The German Challenge

If France in the late eighteenth century was the great centre of mechanistic, 'Enlightenment' science, with its associated materialism, Germany was a country, or collection of states, which had never fallen beneath the spell of Newton and Locke. The great German figure at the beginning of modern science had been Paracelsus, the bombastic assailant of ancient medicine whose 'alchemical' remedies involving inorganic substances made him famous, or infamous. Not only did he introduce mercury for the treatment of syphilis, which swept through Europe in the 1490s, but he also saw everywhere in nature a web of correspondences and mysterious forces. Paracelsian medicine involved astrology as well as alchemy; symbols loaded with magical significance as well as experiments with crucibles and furnaces. Paracelsians expected to find resonances in nature, and rejected clarity if it meant dealing only with the shell and surface of things: their world-view was the antithesis of that of Descartes, who sought for the clear and distinct. In Rosicrucian and other occult writings, and in speculative works, elements of Paracelsian science survived down to the eighteenth century, to be taken up by authors involved in the Romantic Movement.

There was also a strong tradition of more rigorous metaphysics. Leibniz had seen that beneath Newtonian physics lay assumptions about the nature of space and time which he thought false.[1] He also criticized the 'action at a distance' which he perceived in Newton: gravity could not act across void space, and to suggest it could was to propose a perpetual miracle. Atomic theory came in for attack too, for Leibniz found in nature infinite variety rather than endless repetition. He saw gradual transitions everywhere, nature making no leaps, rather than sharp boundaries or discontinuities between things. His controversy over these matters with the Newtonian Samuel Clarke, published post-humously in 1717, was perhaps a draw; most contemporaries were pragmatic enough to accept the Newtonian system because it worked so well, but especially in Germany there were many readers of Leibniz for whom therefore Newton's system could not be more than a provisional scheme.

It was not until the second half of the eighteenth century that Latin ceased to be the learned language for Germans. Then came the work of Herder, one of the first to appreciate the importance of historical context and change through history; Goethe, poet, novelist and man of science; and Kant, astronomer and then philosopher. Herder is of interest in the history of science as one of those who insisted on language as the essence of humanity, separating us from animals. Goethe has a more direct importance, because he used science old and new in his writings, and even wrote some scientific works. His *Faust*, which occupied him on and off for the last sixty years of his life, was based on the story of a Renaissance magus, and used alchemical imagery and elements picked up from a reading of Paracelsus. Alchemy is a confusing and fascinating field just because it is a compound of the psychological and the chemical, aiming at eternal life as much as at imperishable gold, and its literature abounds in symbols and resonances. To use alchemical imagery in literature was not therefore unprecedented, though Goethe's is probably the only extended masterpiece of literature in the genre.

More striking was his use of modern chemical symbolism.[2] About 1600 two Germans, Oswald Croll and Andreas Libavius, had written chemical works: the former's was resonant, poetic and hard to follow except for the adept already instructed in alchemy; and the latter's was straightforward and plain, an early textbook. It was Libavius' ideal of clear language, rather than the more poetic vision of mystery, harmony and unity, which prevailed in chemistry. The eighteenth-century reforms of chemical language carried out by Lavoisier and his associates removed old names with astrological or alchemical overtones, like 'vitriol of Saturn' and 'regulus of antinomy'. Names were to be suggestive only of manifest properties, or of (supposedly) established theories.

One term frowned on by the French but surviving into the nineteenth century was 'elective affinity'.[3] Whereas gravity is a universal attractive force, chemical attraction is not: some substances are highly reactive, others are inert, and in general things seem choosy. In a metaphor deriving from marriage, substances were said to display 'affinities': while blood relations are 'kindred', in-laws are 'affinity' in the tables stating who one may not marry. In his novel *Elective Affinities* (1809) Goethe brought back the idea from chemistry to humanity with the irony of which he was a master.[4] His characters discuss a double decomposition: AB + CD = AD + BC. 'Bring the two pairs into contact; A will fling himself on D, C on B, without its being possible to say which had first left its first connection, or made the first move towards the second.' The reaction between the two pairs of people does not go as had been forecast: whereas the chemical reaction proceeds smoothly to completion, that between the people is disastrous, with death and despair as the consequences.

Because of Goethe's skill, what might have been a mechanical, or chemical, application of a scientific idea does not lead to a dry novel; and one thing which can be learned from it is the inappropriateness of applying models from inorganic nature to mankind, in the manner of the Enlightenment. When it came to anatomy and physiology, Goethe was prepared to look for harmonies and homologies. He discovered in the human skull traces of the intermaxillary bone, prominent in other animals; and in the morphology of plants he traced the development of flowers out of leaves. Goethe believed that men of science had been like little children looking at a mirror and trying to get at what they thought they saw behind it. The way forward was not to try to explain through hypotheses, especially mechanical ones; but to look for primal examples, *Urphänomene*, which like Bacon's 'shining instances' somehow summed up both the concrete instance and the ideal, and to generalize from them. In biology, this led to an emphasis on types or archetypes; and in Goethe's other great field of interest, optics, to seizing on certain experiments which seemed to him crucial and simple.

Goethe's work on colours, *Farbenlehre* (1810), was extremely influential, though not among those actually engaged in work in optics outside Germany; they were busy debating whether a wave or particle model was best, or rather was true. For Goethe, 'we should theorize with mental self-possession, and, to use a bold word, with irony'. He believed that Newton – a name which for him stood for the whole tradition of mathematical physics – had gone off on completely the wrong tack in trying to impose a mathematical model. What was needed was to concentrate on what we actually see. Goethe included illustrations with sharp boundaries of black and white to demonstrate how we may see colours where there are 'really' none, and how different in size a white circle on a black ground looks from a black one on white. He had noticed how red objects in a picture come forward, and blue ones recede; he thus knew that there was more to perspective than geometrical models.

His science was thus closer to psychological than to what we would call physical optics, though this is a distinction which he would not have wished to make. In place of the mechanical science of the Enlightenment he wanted a dynamical one. Mechanics, with its associated particle theories, was a world of endless little things; what made the world interesting was our interaction with it, and it was to be understood in terms of forces. These forces moreover were polar; to each there was an opposite, like the north and south polar forces of magnetism, and the positive and negative of electricity. Indeed, the observation of polar magnetic forces was one of *Urphänomene*. Goethe rejected the Romantic outlook, seeing himself as in the Classical tradition; but his criticisms of the Enlightenment were grist to the mill of Romantic writers of both literature and science.

The same could be said of Kant. With his emphasis upon the way we impose categories on the world, he undermined the idea that we simply observe facts. All our perceptions become to some degree theory-laden for the Kantian. His antinomies again seemed to indicate that scientific ideas pushed too far led to incoherence. But in his writings on teleology he sent the study of living creatures in Germany on a new course.[5] In Britain, the way an organism works as a unity was taken as proof that God had designed it; in Germany, it followed from and vindicated the view that the parts and organs can only be understood when their function is taken into account. An animal is more than the sum of its parts. This functional approach was characteristic especially of the great embryologist Karl von Baer. Working with the newly perfected achromatic microscope (in which doublet lenses of different glasses cut out the coloured fringes round the image),[6] he followed the full development of the mammalian embryo. The evolution of an embryo, as its parts are gradually differentiated, is a field where holism and goal-directedness are obvious. The processes involved eluded the rather primitive chemical and physical explanations available to materialists; and down to the middle of the nineteenth century there was general belief in some kind of vital force, responsible for the evolution of organisms from seed and for maintaining those harmonious functions which constitute life.

The concern with a life force was not confined to Germany, but had many supporters in Britain; but in Germany it went with that dynamical emphasis in science which had the authority of Leibniz, Goethe and Kant.[7] Romantic successors of Kant, especially F. W. J. Schelling, worked out a philosophy of nature, *Naturphilosophie*, which was based upon the notion of polar forces. Schelling's *Ideas towards a Philosophy of Nature* came out in 1797, and his *First Outlines* of the system two years later. He was a Romantic rather than a systematic thinker, and some of these ideas were worked up in a more coherent form by Hegel in his *Encyclopaedia*, which first appeared in 1817.[8] We now tend to suppose that physics is the fundamental science, but to Schelling electricity and chemistry were getting below the surface of things rather than just dealing with 'models' of clocks and billiard-balls. They were therefore the exciting sciences; and indeed not only in Germany but also in France and Britain, Fourcroy and Davy attracted great crowds to lectures on chemistry. This was partly because it was in a state of flux, following on Lavoisier's work, and seemed accessible in a way that highly developed sciences like astronomy were not; but also because it seemed that chemistry might somehow have the key to the universe.

Chemistry with its elective affinities was really a science of forces rather than of matter. And in chemical reactions the clash of opposites resulted not in annihilation but in synthesis: oxygen and hydrogen came together and formed water, themselves ceasing to exist as separate entities. The crude model of the dialectic of Hegel and Marx began in the chemistry of about 1800. The problem was that the elective affinities were very difficult to deal with exactly. In different circumstances, different metals seem more or less reactive than one another; and the displacement reactions by which the strength of affinities could be assessed were not a matter of opposites but of similarities. Thus copper will be displaced from copper sulphate solution if one adds an iron nail to it; chemical elements seemed to lie on a continuum of activity from hydrogen to oxygen rather than at opposite poles.

Nevertheless, Schelling postulated a world in which the various polar opposites that we see are all expressions of one underlying force. What were about 1800 generally supposed to be weightless fluids – heat, light, electricity and magnetism – were all seen as effects of this universal force. And just at this period came the first interconversions of forces which gave an empirical justification to Schelling's views. In 1800 A. Volta published his discovery that when different metals are dipped into (impure) water and connected, an electric current flows. Volta interpreted this as due to mere contact, but contemporaries such as J. W. Ritter, associated with Schelling, thought it to be produced by the chemical action of the water on one of the metals. The system was soon reversed, as in Davy's work, so that instead of a chemical reaction generating electricity, an electric current was used to make a reaction go: first the decomposition of water, and then of potash, soda and other compounds.

Davy could therefore conclude that electricity and chemical affinity were manifestations of one force; he seemed to have proved it experimentally, some ten years after Schelling had asserted it. Other interconversions or correlations were the discovery of heating rays beyond the visible spectrum, proving the analogy of heat and light, by William Herschel; and then the discovery of chemically active rays at the other end of the spectrum, the ultra-violet, by W. H. Wollaston in England and by Ritter independently. The great triumph came later, in 1820, when Hans Christian Oersted – a friend of Ritter – at last succeeded in demonstrating that an electric current could produce magnetic effects, as he had been predicting for some years in works his contemporaries regarded as 'metaphysical'.

Ampère in France, Joseph Henry in the USA and Faraday in Britain took up Oersted's work and electromagnetism soon became international, its discoverer playing little part in its history thereafter. But Oersted's essays and addresses on various occasions were put into a book, *The Soul in Nature* (1852) which gives us a picture of how he thought. He believed that our reason was in union with the divine reason, and that we could therefore hope to understand the world rationally rather than merely experimentally, to see why things must be the way they are. He saw flux and process everywhere: apparent rest is really dynamic equilibrium; and people only endure, like waterfalls, through the ceaseless change of the matter which composes them, and which is to be understood not as brute inert substance but as an expression of activity. He suggested a theory of development, according to which the solar system had evolved according to law, and the metals were 'similar resting-points in the development of the earth, and that there might exist a similarity in the laws by which both developments have taken place'. Above all he urged that science was enobling; that utility was not enough: the scientist should be a sage.

We have included the Dane, Oersted, among the Germans because of his connections with Germans and his use of their language in some at least of his papers, though his discovery of 1820 was circulated in Latin! Eighteenth-century Germany had been a mass of states of various sizes, owing some kind of allegiance to the Holy Roman Emperor; but this ramshackle arrangement, really united only by language and culture, was swept away in the French Revolutionary and Napoleonic invasions. These in turn produced a patriotic reaction; and after 1814 some of the smallest states, free cities or prince-bishoprics, disappeared, swallowed up into states of which the populations now began to feel themselves German. There were advantages in the political fragmentation which endured down to 1870, when Bismarck incorporated all but Austria-Hungary, by then the weaker and humiliated state, into a German Empire under Prussian domination. With several states, there was competition not merely over opera-houses but also in universities; and there were posts in the various courts open to men like Goethe. Whereas in France all intellectual activity became concentrated in

Paris, in Germany there was no one capital, and culture was more widely diffused. The USA at the same period had no one intellectual capital either, and reaped some of the same advantages, though in nineteenth-century science it was never more than a second-class power.[9]

Prussia played an important part in the defeat of Napoleon, culminating in Blücher's crucial arrival at the battle of Waterloo. Among the leading statesmen of Prussia was William von Humboldt, brother of the explorer, and both a linguist (with particular interest in the Basque language) and political philosopher.[10] It was he who planned the University of Berlin, which is now named after him, and which became the leading institution in Germany. In the eighteenth century, the German pedant could be a figure of fun; but even then much of Germany was, like Scotland, a relatively poor country with a relatively good educational system.[11] In the later eighteenth century the most eminent university was Göttingen, in the Hanoverian dominions; and there research was done by some professors, notably Blumenbach in anatomy and physiology. But it was Humboldt who made prominent the two ideals of the university, *Wissenschaft* or knowledge for its own sake (and not mere pedantry), and *Bildung*, or intellectual and moral self-development.[12]

These ideals were far from the utilitarian emphases so often invoked by those who favour more science in universities. Science in the first decades of the nineteenth century was taught in the philosophy faculty of German universities, and this reinforced the idea that it was as much in the business of understanding the world as of changing it. But although Germans were appalled at the utilitarian attitude to science they perceived in Britain, they had themselves pioneered a vocational system of chemical education which in time led to the formation of great research schools, and then to a dominant position in chemical industry.

The German chemical community had been strongly associated with the old theory, and broke up under the impact of French theory and French troops.[13] But with the rise in Prussian bureaucracy and Teutonic efficiency in the Napoleonic years came legislation to ensure that pharmacists received formal training before they were allowed to practise. They had been taught by apprenticeship, and counted as the lowest of the 'learned' in eighteenth-century Germany; formal training raised their status, and presumably also their competence. Other German states soon followed Prussia – it was convenient for everyone if there were common qualifications – and the formal training gradually became longer.

Justus Liebig, after studying chemistry at Erlangen, went to Paris, where through the influence of Alexander von Humboldt he worked with Gay-Lussac. In 1825 he was appointed to the chair of chemistry at the tiny university of Giessen.[14] There he built up a research school, based on his methods which made organic analysis a matter of routine. His was not the first teaching laboratory, and he did not even invent the Liebig condenser, with its jacket of flowing cold water to trap volatile distillates; but his was the first nursery in which research chemists were bred in large numbers, and it became the ancestor of the modern chemistry department. It had begun on a basis of training pharmacists, but Liebig interested them in research and those trained with him found that their skills might lead them instead into academic life or into chemical manufacturing. There had been a teaching laboratory at Glasgow, under Thomas Thomson, before Liebig's; but Thomson seems to have given his students less freedom to work on their research rather than his, and the system never took off as Liebig's did. Thomson also had given up control of his journal, *Annals of Philosophy*, soon after being appointed to the chair at Glasgow in 1818, whereas Liebig saw the importance of a professor controlling an outlet for his students to publish in.

In the 1820s there began the meetings of German men of science, held annually and each year in a different place; this provision was a consequence of the political fragmentation of Germany. The moving spirit of this organization was Lorenz Oken, the great exponent of *Naturphilosophie* in the life sciences. His *Physiophilosophie*, written in 1810 'in a kind of inspiration' and therefore 'not so well arranged as a systematic work ought to be', was meant to cover the whole of nature. It begins with the conception of science, and goes on through astronomy and physics up to botany, zoology and finally psychology; everywhere there are analogies, parallels and polarities, and the book is organized as a series of numbered paragraphs, many of them gnomic. Then and now, some readers will find remarks such as 'the elements are only gradations of light, colours', or 'the nest of a bird is a spiritual repetition of its plumage', curious and original, while to others they will be mumbo-jumbo. To Liebig, this kind of thing was the Black Death of the nineteenth century; in Germany one could find both speculative insights and also highly empirical research and teaching.

Naturphilosophie, with its emphasis on polarities and on analogies running right across all the kingdoms of nature, and on the power of the mind to anticipate experiment or observation, was distinct from the Kantian teleology of men like Baer, though this was not always obvious to contemporaries or to later critics. Oken and those like him saw a biology of types, sometimes real creatures but often ideal and therefore really archetypes.[15] All crustaceans might be seen as exemplifications, for example, of a single type; all fell short of this Platonic ideal in this case, for none had the full complement of appendages. The crabs, at least in adulthood, departed more from the type than lobsters, and could be seen as higher in the group. Study of embryos could be a guide to an animal's place in the scheme of things; and the term 'evolution', which was used of the growth of an embryo, might be applied to the ideal evolution of the various members of the groups from their type. Thus while in Britain and in France there were definite, if not really testable, theories of development in the early nineteenth century, in Germany there was a transcendental theory. Darwin's version of 1859, with its hit-and-miss mechanism of natural selection, was rather different from any of those produced half a century or more before.[16]

Also in 1810, Samuel Hahnemann published the first edition of his *Organon of Medicine*, the basis of homoeopathy, an 'alternative' system of medicine which still survives.[17] He was highly critical of the heroic medicine which had been favoured by Paracelsus with his powerful drugs, and of the medicine of his day, which he saw as directed merely to the palliation of symptoms. He believed that illnesses were not caused by 'fictitious' disease entities, something material to be expelled; what had happened was that some dynamic affection was causing symptoms, and that balance would be restored by administering minute doses of a medicine which would produce those symptoms, but more strongly. It was vital to grind up the medicine very fine, and give tiny doses: matter when finely divided and successively dissolved displays its hidden forces, the material being spiritualized. Certainly the medicine of Hahnemann's day probably killed as many as it cured, and at worst his tiny doses could have done little harm compared to bleedings and purges. To our eye the book is a curious mixture of clinical observation and rather strange analogy of disease and vital forces with electricity and magnetism; but these are simply a reflection of the scientific milieu in which the system was born.

Meanwhile at Göttingen one of the greatest mathematicians ever, Karl Gauss, was doing work comparable with the best being produced by the Parisian school. He worked out the method of least squares, and the Gaussian distribution, in the course of work on errors in astronomical observations; pioneered non-Euclidean geometry; worked on number theory, geomagnetism and electromagnetism; and in general played a major part in that mathematical physics which in the later nineteenth century would have a prominent German contribution. In this Alexander von Humboldt also had a complementary part to play. He was almost the founder of modern geography, and as befitted one from Germany, a congeries of states, who had worked under French auspices, he was prominent in working for international cooperation in science. Networks of geodetic observatories, and series of expeditions and surveys, were the fruit of this work; and indeed the 'big science', that which received government subsidy, of the 1820s and 1830s had been characterized as Humboldtian.[18] This was a descriptive kind of geophysics. Humboldt was a powerful lecturer and writer, and almost straddled the arts/sciences divide: his *Aspects of Nature* has some magnificent set-piece descriptions of scenery through which the botanist and geologist also shows, and his *Cosmos* was an extended effort (begun when he was seventy-six) to synthesize a Romantic sensibility and an empirical attitude. This 'great work of our age' was never of great influence: its author, who saw himself always as 'a man of 1789' when at nineteen he had hurried to Paris in the Revolution, lived on to 1859, when science had become increasingly specialized, professional and mathematical, and his spirit had ceased to guide it.

The range of German thought and practice made Germany attractive to those in Britain for whom a Baconian science and a Paleyan natural theology were unsatisfying. One of the first of these was S. T. Coleridge, who had read widely in Platonic and neoplatonic philosophy, and whose enthusiasm for Germany was kindled by Thomas Beddoes, Davy's patron and himself one of the few in Britain to recognize the importance of Kant, whose work he reviewed (none too favourably). Coleridge went to Germany, heard Blumenbach and read various authors, and came back moderately fluent in the language; whereupon Beddoes did him another good turn in introducing him to Davy, and a bad one in prescribing him opium for toothache. Coleridge came back filled with enthusiasm for a world of polar forces,[19] and no doubt communicated some of his excitement to Davy. Later in print he praised Davy and others for bringing a dynamical spirit into science: for he realized the difference between speculating that all force was one, and proving that chemical affinity was electrical.

Davy on the other hand never read German, and when in his last years, a lonely and prematurely old man in his late forties, he went fishing in Carinthia, he carried about with him a few useful phrases written out by the innkeeper's daughter. Nor did Faraday read German: it was a language little known in Britain in the early nineteenth century. When Pusey went to Germany in the 1820s it was said that there were only two others in Oxford who knew the language; German scholarship whether in sciences or humanities was hardly known, and might almost have been going on on the Moon. French and Italian were the foreign languages that were easily read, and every gentleman would expect to manage Latin. Some ideas from *Naturphilosophie* came through via the French, for example in Cuvier's *Rapport Historique* of 1810, which was a review of the progress of science since 1789; while it was hostile to speculative German thought, at least it mentioned it.

It is tempting to make out connections between the belief that all force is one and the later doctrine of the conservation of energy, and indeed there do seem to be links.[20] How strong these links were in Britain, however, it is hard to say. It seems that rather as Coleridge went to Germany with his head full of

Proclus and Plotinus and then grafted on some Schelling, so Davy and his contemporaries had a Newtonian vision of all attractive forces being one which chimed in with what they heard from Coleridge. Oersted's discovery was taken up by Faraday, but again direct influence of Oersted's metaphysics is harder to establish; Faraday certainly became committed to an increasingly dynamic world-view, where forces were much more important than matter.

Goethe's *Theory of Colours* was translated in 1840, with some of the more polemical parts left out, by Charles Eastlake. Eastlake was both a Royal Academician and a Fellow of the Royal Society, but it was in his capacity as an artist that he seems to have been most interested in the book, and it seems to have been regarded as no more than a curiosity by men of science. This was partly because it was then thirty years old, and therefore out of date: attention had shifted by 1840 to details of the wave theory of light, with the transverse waves of Fresnel and Arago prevailing over the corpuscular theory. The questions being asked related to the luminiferous ether and interference of waves, and the idea of abandoning a mathematical approach to optics would have seemed ludicrous. By the second half of the nineteenth century, the wave theory of light was being used as an example of a mathematical construction which clearly did fit nature. Outside Germany, Goethe's theory was seen as an aberration or an awful warning.

Oken's *Physiophilosophy* was translated for the Ray society, which was founded in order to publish works of natural history; some of its earliest books were translations from German, and Oken's appeared in 1847. Once again, it must have seemed out of date to some members of the society; but Oken's name commanded some prestige, and there was widespread interest in analogies, affinities, homologies and types. Edward Forbes, the great hope of British palaeontology at the time, wrote in a letter of 1847 that Oken was one of those who perceived analogies more readily than affinities; but that Goethe – with Aristotle, Linnaeus and Brown – combined 'an apparently intuitive (really logical) perception of analogy with the power of accurate recognition of affinity, and both with the intuitive perception of ideal form, = beauty. The greatest leading minds belong to this class'. In the days before there was a respectable theory of development, perception of analogies, or distant resemblances, was necessary for getting clear groups of plants or animals into a whole pattern. Oken's synthesis therefore was still timely; he wrote a preface to the translation, noting that all the parallels had not been worked out, that there was something to change every year, and that 'the present is but a sample of how we are to proceed in our desire of obtaining a Natural system'.

Just a few years later, in 1850, the *Researches* of Karl von Reichenbach were translated into English. They were concerned with 'magnetism, electricity, heat, light, crystallization, and chemical attraction, in their relations to the vital force', and had appeared in German from 1845. Reichenbach was a chemist, who had improved iron manufacture in Austria and had done research on tar; his work appeared first in Liebig's *Annalen*, and received the cautious encouragement of Berzelius. It was assailed by the physiologists Johannes Müller and Du Bois-Reymond as hopelessly unscientific, but it proposed a new kind of 'fluid, or imponderable, or dynamide, or influence (whatever name be given to it), which, by means of *passes*, of *laying on of hands*, and of *transference* or *communication*, produces astounding physical and physiological effects.' This force he called *odyle*, and it was somehow magnetic, and could be seen by sensitive persons as a kind of aura of light. The first sensitives were sickly young women, but later experiments showed that there were those without these disadvantages who also saw the effects, and indeed Reichenbach thought that about one-third of mankind was sensitive.

Sensitive people saw something like flames spreading from magnets, and could even blow them about; the north pole might be enveloped in blue flame, and the south in red. The aurora borealis, known to have some connection with magnetism, might be the odylic flames of the Earth's magnetic field; and odyle also seemed to display polarity like electricity and magnetism. The light that hovers over graves was no doubt odylic; the new force embraced the whole universe. The book, which followed a brief account in English by the translator which had caused great interest, is full of case studies; flames were sometimes seen as high as a man, though some effects were only detectable when the observer was pregnant. We shall bear Reichenbach in mind when looking at scientific scepticism and credulity in the next chapter; the interesting point here is that the theoretical background to the book is polar forces, convertable one into another.

It is curious that Liebig should have published material of this sort, but he must have been interested by the many case studies involved; he had himself written in an amusingly sceptical tone about reports of the spontaneous combustion of drunkards. It was Liebig's translator, William Gregory (Professor of Chemistry at Edinburgh), who also translated Reichenbach. Liebig built up good relations with the British Association for the Advancement of Science, and established a very high reputation for himself in Britain, especially with his *Chemistry in its applications to Agriculture and Physiology* (1840), *Animal Chemistry* (1842) and *Familiar Letters on Chemistry* (1843). In the hungry forties, the hope of Davy, Johnston and others that agriculture might be made really scientific and the land much more productive seemed really worth pursuing; and Liebig's simplistic methods, relying heavily on inorganic manures made not too soluble, were attractive. In the event, the improving landowners of the 1840s proved as cautiously empirical towards Liebig as their fathers had been towards Davy's *Agricultural Chemistry* of 1813; but Davy's approach had been different, in that instead of starting with analyses and boldly reasoning from them, he had cautiously used chemistry to vindicate the best current practice.

Liebig's *Agriculture* was translated not by Gregory but by Lyon Playfair, who had studied at Giessen and was to succeed Gregory at Edinburgh. But he is of much greater importance as a statesman of science. Indeed Playfair was one of the great proponents of applied science in nineteenth-century Britain, playing a major part in the Great Exhibition of 1851 and thereafter holding important political office and serving on many commissions concerned with science and education. By 1840 Germany was seen as the centre of activity in chemistry, and anyone with any ambition to get on in the science went like Playfair to a German university to do a PhD; this degree was not then available in Britain, where doctorates were awarded in law, theology and medicine, and generally in the first two cases for published work rather than for a thesis. By this time, then, ignorance of German thought was no longer a feature of chemists. Outside their ranks, however, there were many like Darwin who struggled rather hopelessly with the language, and one of his friends began it several times.

Darwin had read Humboldt's *Personal Narrative* in English translation (1814–29) before going on his own voyage to South America; and there were rival translations of the *Cosmos* and the *Aspects of Nature* in the 1840s and 1850s, so that his work was widely available. The speculative and the empirical traditions in German science were both therefore represented, along with Humboldt's writings, which in some sense were a synthesis of both. But the more technical German science was patchily represented. The Ray Society published some natural history, and chemical work was translated – not only Liebig, but also Leopold Gmelin, whose *Handbook*, a systematic account of the whole science, appeared in English in 1848–52; it then filled six large volumes, and as an invaluable work of reference it has gone on being revised and expanded up to the present.

Otherwise papers might appear in German periodicals and remain quite unknown to men of science in Britain; and to a lesser extent the same was true of work published in French and Italian. In 1837 the enterprising scientific publisher Richard Taylor brought out the first volume of a new journal, *Scientific Memoirs*.[21] This was to print translated papers from foreign journals exclusively and Taylor arranged for distributors in the USA as well as in the British Isles. In his preface written on the completion of the first volume, Taylor stated that the publication had sold only 250 copies and was not covering its cost; publishing science in the nineteenth century was not usually a way of printing money, although one could later do very well out of textbooks.

The journal was very miscellaneous in its coverage, and the authors translated include some now very famous names like Ohm, and others now forgotten. Taylor depended upon distinguished scientists to suggest papers which should be translated, and they range from natural history to thermodynamics. After the first volume, there was a gap and the second came out only when support had been forthcoming from the BAAS; it included much material on terrestrial magnetism by Gauss and by Wilhelm Weber, as well as Ohm's paper, and was almost exclusively translated from German. By volume 3, 1843, there were papers connected with early photography; volume 4, 1846, had two papers by the elderly Berzelius, one of them his famous assault upon Dumas for his theory of chemical types, an attempt to explain how the substitution of the very different element chlorine for hydrogen in an organic compound made very little difference to its properties. With volume 5, 1852, Taylor gave up the editorship; and in the following year the last volume appeared, divided into two parts – natural philosophy and natural history. The former had important writings of Clausius and Helmholtz, on energy conservation; and the latter translated some of Baer's work. There the series stopped; but its work was perhaps done, for Anglo-German contacts by the 1850s were close as far as science was concerned and those of T. H. Huxley's generation looked to Germany with a certain longing as a country where science was really respected and rewarded, and whence advances in fields like physiology might be expected.[22] Not only Baer, but also Müller and Du Bois-Reymond were behind the new physiology which Huxley and his disciples brought to Britain.[23]

At the opening of the century, then, German thought was particularly attractive to those who liked metaphysics and speculation. By the mid-century one could find there not only odyle but also the latest empirical and mathematical work right across the sciences. French dominance was a thing of the past, though naturally in particular fields the French might be leaders. What seemed to be needed in science by the middle of the century was a certain amount of bold speculation, combined with rigorous testing. To be successful in science, one needed imagination: the balance looked more Germanic than Baconian.

The problem is that imagination in science can easily lead astray. The light that gleams upon the benighted traveller may be the flicker of the idiot's lantern drawing him further into the boggy wilderness. The wild analogies of *Naturphilosophie* are not so different from conjectures about conservation of energy, or of homologies in the animal kingdom, and indeed it is plausible to say that we could not have had one without the other. What is striking about scientific communities is their scepticism. The person who proposes a new interpretation is confronted with disbelief, or perhaps indifference. This is an essential characteristic of science; we are surrounded by a booming, buzzing confusion, and we have to select what we consider relevant – there is not much room for the really unexpected. Men like Darwin and his allies had to struggle to get their ideas across, and there has to be an element of rhetoric about all scientific publications.

It is supposed that this scepticism will be a filter, allowing through what is good (probably after a little delay) and stopping what is bad, or pseudo-science. This does not always happen: fashion in science may sometimes indicate progress in understanding, but may be sometimes mere fashion[24] – as we know in our century from notions of diet and health. The sceptical nineteenth century was also a time of great credulity; and we can learn much about science from looking at those bodies of knowledge which did not acquire the longed-for scientific status. Our next chapter will be concerned with arguing with sceptics.

5

Arguing with Sceptics

Science is, and always was, based on a judicious mixture of empiricism and faith. These things can degenerate easily into gritty factuality, dogmatism or credulity. Baconian philosophy gave the Royal Society its motto, *Nullius in verba*, rely on nobody's words; but Bacon himself went beyond facts in his inference (very shakily supported in his own day) that motion was the cause of heat. Some of Bacon's contemporaries were scornful of the fringe medicine of the early seventeenth century, with its astrological connections and its claims to cure at a distance; and were also sceptical about witch-craft and the alleged activities of ghosts and spirits, though there were prominent early Fellows of the Royal Society who passionately believed in such things. Men of science were also sceptical about the notion that fossils were really the remains of extinct creatures, and that meteorites had fallen from the sky; it was and is difficult to know where to draw the line. Nature is always capable of surprising us, and a dogmatic scepticism is hopeless as a guide to the advancement of learning, or the finding of more things than we had dreamed of.

This makes the history of sciences which failed to become established, or which died, particularly fascinating. In arguing for or against some new conception of the world, men of science disclose their assumptions about the way things are, which without a crisis they would have kept to themselves; and people are forced to spell out what in the ordinary way they take for granted. It is helpful too, if we are to understand the innovations of a Cuvier, a Helmholtz or a Darwin, to look at some would-be revolutionaries who may turn out, like many alchemists, to be tricksters or quacks playing on the credulity of the public, but who may, and indeed usually did, believe that they were on to the real key to the universe. To use the term pseudo-science is to imply that one is describing impostors and charlatans; it generally involves wisdom after the event, which is easy and cheap. The task of the historian is not to cast moral opprobrium but to understand.

One of the great hopes of the early nineteenth century was that a science of man was just around the corner. The first hope was in physiognomy, the idea that the face properly understood would reveal the character and the abilities of the person.[1] This was probably more hopeful than palm-reading, and we all know people with sad or happy, or with hard or kind faces. John Caspar Lavater, a Swiss pastor of the late eighteenth century, interested in mesmerism, became famous for his 'physiognomical cabinet' of portraits with commentaries; he eventually published his *Essays on Physiognomy* (translated by the atheist Thomas Holcroft), which circulated widely and long – the tenth English edition came out in 1858. The book includes hundreds of portraits, often of the eminent (of various degrees of authenticity), and some portraits of animals, birds, insects and fish for comparison. He was also interested in skulls, and included plates of various ones; among them, as became usual in the nineteenth century, that of a male German, a typical European skull, was claimed to show itself fitted for 'more delicate and spiritual enjoyment' than those of other races, although it had come from one who while not stupid was not a man of genius.

Lavater recommended his work to princes and judges, and indeed it would be valuable if character could be as easily read as he implied. It should perhaps be recorded that he met his death (rather like Archimedes) when Zurich was occupied by the French, unfortunately misreading the temperament of some soldiers. His ideas had little scientific basis, but were soon absorbed within a system which really promised strict predictions: the phrenology of Gall and Spurzheim.[2] The basis of this science was the idea that the various functions of the brain were located at various places within it, and that if these functions were well developed then that portion of the brain would be correspondingly large. In order to accommodate it, the skull must therefore bulge at that point. The shape of the skull hence indicated the shape of the brain, and thus the faculties of the person investigated. Examining the skulls of the benevolent, the destructive, the amorous, the secretive, the hopeful and the orderly, phrenologists hoped to identify the bumps that went with these various qualities. If only the science were made precise, then the task not only of the judge but also of the teacher would be made simpler. We make our faces, to some degree at least, but hardly our skulls unless we belong to groups which use cradle-boards; so feeling heads is a way to find out innate tendencies. The teacher who determined the bumps in his class would know just what qualities to encourage and which to repress in the effort to produce well-adjusted adults.[3]

Phrenology aroused considerable passions, for and against; scepticism was not confined to the scientific community, and some prominent medical men as well as various artists took up the ideas – at least as working hypotheses in the one case, and as a guide to plausible heads and faces for historical characters in the other. Indeed, some phrenology like some geology and meteorology came to be seen as important for artists in the scientific age. On the other hand, the critic William Hazlitt in 1829 wrote one of his entertaining essays on 'phrenological fallacies', seeing the science as the 'grave or epitaph of the understanding'. The phrenologist for him distracted attention from the most important parts of the head, that is the face, in favour of the cranium; more seriously, the idea of numerous separate faculties was at odds with what we know of the great, who differ from the ordinary not in one organ but in the whole mind. The moral capacity could not be cut up into 'wiredrawn departments', and phrenology eluded common sense.

Phrenologists naturally saw things rather differently, and in 1823 in Edinburgh the first volume of the *Phrenological Journal and Miscellany* began to appear, in quarterly parts. Edinburgh indeed became the centre of phrenological argument, for and against, no doubt in part because of its pre-eminent medical school. The *Journal* is rather different from most scientific publications because it is highly argumentative in tone, making no attempt at cool detachment – which helps to make it readable.

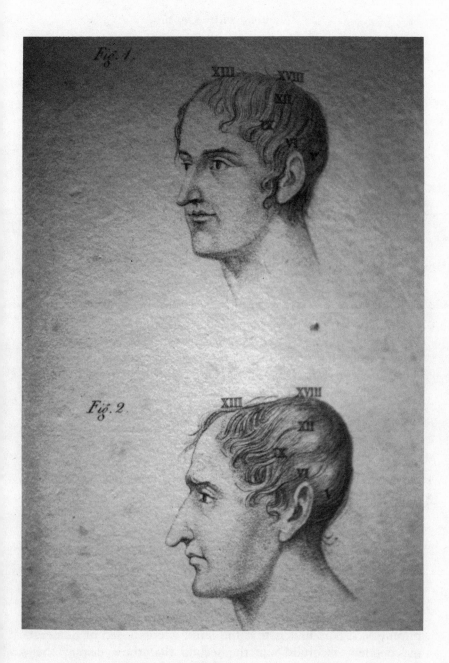

Figure 3 *Phrenological heads from J. G. Spurzheim,* The Physiognomical System, *second edition, 1815, plate XI.*

It begins with a polemical Introductory Statement attacking the various publications which had made their ignorant and coarse assaults upon the science, and concludes with a warning:

> One sign of the times is worth the regard of the most securely established philosopher. The doctrine, which he unwisely despises, is rapidly taking hold of the reason, and delighting the imagination of the rising generation. *They* have no exclusive theories which they love better than truth; no philosphic dignities and reputations in jeopardy; no pride to be offended by the success of a system which they have not committed themselves by contemning.

These youths will soon be men, and will look back to *Alma Mater* with contempt, sending their sons to the schools of the phrenologists instead.

The University of Edinburgh survived, but the argument was one used by Lavoisier, Darwin and Planck to rally their supporters in the face of conservatism and scepticism. Davy's discovery of potassium, and his work on chlorine, were only slowly accepted; the Church Scientific, as T. H. Huxley called the scientific establishment, moves cautiously, and the eminent are not delighted to see their favourite views falsified. Because there is an element of faith in all science, falsification or justification can never be complete; there comes a point where the most plausible or the simplest explanation is to be chosen, but these things are in part a matter of taste. It is usually but not always the case that the young go first for the new theory; partly they do so because it may be more fertile in research problems.

The *Journal* contains not only descriptions of the heads of heroes and villains, but also some phrenological criticism of Shakespeare and other eminent poets. It is one thing from doubtful portraits or busts of the illustrious dead to reconstruct their characters, but for Shakespeare's Macbeth there never was a head to measure. The point is that there are two aspects to phrenology. The first is that the mind is composed of numerous independent faculties, and the second that these display themselves as bulges in the brain and hence as bumps in the skull. Phrenological criticism of poems and novels emphasized how great authors in portraying character had made use of the categories of the phrenologists, and shown us magnificient examples of the predominance of various faculties. In this, phrenology is not unlike Darwinism, which involved a general principle of development and a specific mechanism of natural selection. Thus any evidence of evolution gave some support to Darwinism, while phrenologists reckoned that any evidence of separate faculties supported them.

The *Journal* also reviewed other works not dealing directly with its subject, such as John Franklin's *Journey to the Polar Sea* (1823). This was a desperate enterprise in northern Canada, involving travelling down the Coppermine River and navigating the Arctic Ocean in birchbark canoes; only a few of the heroic participants survived the return trip to civilization after eating their boots and at last being succoured by Indians. The reviewer did not concern himself with whether Franklin's head indicated that he possessed enough caution and prudence, but more in sorrow than in anger he complained that the members of the party were untrained in phrenology and that a splendid ethnographical opportunity had therefore been missed. There were indeed valuable reports on various Indian and Eskimo groups, but no attempt had been made to correlate them with bumps and thus science had been less advanced than it might have been.

One member of the expedition, John Hepburn, an ordinary seaman rather than an officer, had allowed his head to be charted, so revealing that his bumps had been in accordance with his excellent character. Other papers in the journal duly sought to demonstrate close ties between character traits and shapes of heads, and to deal with those who thought that because a murderer happened to have a particularly large bump of benevolence, phrenology was refuted. One faculty did not make up a character; some thirty had been identified already, and it was the particular mix of them, and the way they had been developed during life, which lay behind our actions. This kind of admission had the awkward result of making phrenology in effect unverifiable, because it no longer generated simple predictions which could be tested; this was one of the things which annoyed Hazlitt, and which makes phrenology not unlike psychoanalysis. Phrenologists nevertheless persisted with their inductive search for correlations, and papers in the *Journal* show how a voyage could be made less tedious by character-studies on the Captain, and an innkeeper's temperament assessed by judicious scrutiny of his cranium across the bar.

Once phrenology was empirically justified by numerous inductions – and its proponents, especially Andrew and George Combe, believed that it was – then the question arose, how it should modify the general world-view. Just at this time the surgeon William Lawrence had caused a scandal with the publication in London in 1819 of his lectures on the *Natural History of Man*. This book, which relied on recent French work to a greater degree than its critics liked, was held to be materialistic, reducing man to a mere animal; and Lawrence duly withdrew it. Curiously, it was ruled to be blasphemous by Eldon, the Lord Chancellor, which meant that Lawrence could derive no profits from it but was also debarred from stopping pirated editions of it; the idea was that there was no copyright in blasphemy. So the work became well-known in unauthorized editions; and while Lawrence's own career was highly successful – he was knighted and became Surgeon to Queen Victoria – the accusation of materialism which seemed to be appropriate to Frenchified physiology was also extended to phrenology. The science needed protection from clerics and metaphysicians; and indeed the charge of materialism was a serious one at this time,[4] for the aversion to what was seen as a doctrine behind the libertinism of the French Revolution was general in Britain.

While physiologists might seem to portray people as no more than machines or chemical laboratories, phrenologists appeared to be determinists. Whereas for astrologers it was planets under which one was born which determined one's character, for the phrenologists it was the shape of the brain. Both systems seemed to undermine freedom of choice and hence responsibility. Phrenologists therefore argued as astrologers had done that they were describing tendencies only; indeed they saw a cranial examination in much the same way as we might see an intelligence or aptitude test. If the bumps merely indicate tendencies, then once again the science becomes untestable; but at least one might allow phrenologists to escape from the charge of rigorous determinism while agreeing that they had indeed displayed good correlations. Materialism was also a difficult charge, because the basis of the science was that mental characters such as benevolence were supposed to be based on the size of a material organ, the brain. The bigger the brain, the stronger the mind; and therefore one might go the whole hog with the materialists, and cross out the word 'mind' wherever one came upon it, writing 'brain' instead. Phrenologists were going along the road which led to the statement that whereas the kidneys secrete urine, the brain secretes thought.

It is therefore not surprising that we find a paper in the journal urging that phrenology and Christianity properly understood are wholly compatible; and another, set out as a dialogue between a phrenologist and a philosopher, to show how well they agree, or should agree. We also find some lighter pieces: a letter from Miss Cordelia Heartless, whose love-life has been complicated by her discovery that it is the brain rather than the heart which is the seat of the passions; a parody of a dialogue from *Blackwood's* which is rather well done; and another parody, of some antiphrenological lectures, put into the mouth of Sieur Donnerblitzen – which is bold stuff from the upholders of a science founded by two learned Germans.

Finally, one might expect that a science in which so much emphasis is placed upon the way the form of the brain affects character would have some social implications. These are illustrated in an essay on the views of Robert Owen the socialist. Abram Combe, the brother of the phrenologists, had become an ardent disciple of Owen. Owenite emphasis on nurture clashed with phrenological doctrine on nature as crucial in human development. It was important to phrenologists to demonstrate the prevalance of, for example, destructiveness as incurably a part of human character, dominant in some individuals, whereas Owen believed that social virtue could be inculcated in all. Anybody who supposes that vandalism, for example, is a new problem should consult the *Journal*, which can function as a source for social history. The review of Owen contains an examination of Owen's head, as furnishing a key to his views: there was a big bump of Benevolence, and of Love of Approbation; but small ones for Destructiveness and Causality, and so on. This seems an unfair game to play on opponents; but the *Journal* gave in this case a right of reply to an Owenite, who appended footnotes critical of the text, making this article a real debate.

Despite all this rhetoric of justification, phrenology never caught on in the scientific community: china heads marked with the bumps were made in large numbers, and the science enjoyed popular esteem while mandarins ignored it. In this it was not unlike evolution in the years before 1859 when Darwin made it respectable. But lest we should suppose that eminent scientists were generally sceptical, we should remember that at much the same period there was great controversy about plurality of worlds.[5] This was the idea that the Earth cannot be unique in having rational inhabitants, but that other planets also must be populated; and ever since this notion had been put about in the seventeenth century by eminent authors such as Kepler, Fontenelle and Huygens, it had become a general tenet amongst men of science.

Its attraction was its appeal to the uniformity of nature. The same laws that regulate the fall of objects on Earth had been shown to govern the planets in their orbits, and at the end of the eighteenth century William Herschel had demonstrated that double stars too move about each other in ellipses under gravity. This doctrine of plurality of worlds was perhaps the way in which the idea of uniformity was brought home to readers, as astronomers cleared their own ideas by writing science fiction. In Huygens' version, the hypothetical planets which he believed to circle the fixed stars were peopled by beings who like us had hands, enjoyed amours and shows, and would know Euclidean geometry; if they were very different from us, it would not be fair.

This fairness was the other great attraction of the idea. First of all, why should the insignificant little Earth happen to have been chosen to have such wonderful creatures as us on it, and nowhere else have such luck? Then secondly, what would be the point of all those fixed stars which could only be seen through a telescope, unless they had inhabited planets, whose dwellers could bask in the light and heat which would otherwise be wasted? Beliefs in uniformity and in teleology thus combined to make plurality of worlds seem a natural and almost axiomatic belief. As late as 1908, Percival Lowell published *Mars as the Abode of Life*. Some writers of the nineteenth century like Davy toyed with the idea that the various heavenly bodies might really be the abode of souls in their pilgrimage from Earth to eternal life.

The weakness and danger of this kind of belief was seen by William Whewell, who possessed one of the sharpest minds in early Victorian England despite being seen as a know-all.[6] In an anonymous criticism (1853) of the idea of plurality of worlds, of which his authorship (as usual at the time) was an open secret, he urged that there was no reason for supposing that the Earth was not unique. Many stars, unlike the Sun, were members of a pair; and there was no evidence for inhabitants either on any known planets, or on hypothetical planets which might be supposed to circle fixed stars. He argued also that the world is full of examples of apparent uselessness: seeds which never come to fruition vastly outnumber those that do; most offspring of animals never come to maturity; 'the vitality which is frustrated is far more copious than the vitality which is consummated'. We should not be too ready to suppose that everywhere which could be a theatre of life is in fact one; and planets anyway have no analogy with seeds or embryos.

In the following year, Whewell's book was attacked by David Brewster (a Scot who was a scourge for all Cambridge men) in a book proclaiming that plurality of worlds was not merely the creed of the philosopher, but the hope of the Christian.[7] It seems curious that there should have been a controversy over what was completely hypothetical; and after all, all that the space flights have shown so far is that there seems to be nobody on the Moon, and no Martians or Venusians for us to talk to. The point was that a dispute about plurality of worlds in space raised questions about plurality of worlds in time.

This form of words was used in the title of James Parkinson's book on fossils, *Organic Remains of a Former World*, 1804–11. This book (which we have already met) is chiefly famous for its illustrations, the text being some way behind the work of Cuvier and his associates; but the book was based upon the idea (found also in Cuvier) that there had been a series of creations of new faunas. Different epochs had come to an end, probably in catastrophe, and a new state of affairs had begun. Only with the existing fauna and flora, the present world, had man appeared.

The view seemed in accordance with what was known in the early nineteenth century, and also seemed compatible with the Bible if it were assumed that Moses had been concerned only with the history of mankind, and not with informing us about ammonites or mastodons.

As further discoveries of fossils revealed vast tracts of time before man had appeared, it became evident that human history was a very small fraction of the Earth's history: for tens of millions of years the planet had been without those rational inhabitants who constituted its glory. It could be supposed that in those ages it was prepared for us, deposits of coal being laid down, and so on; and it might have been that for much of the time it would have been too hot, for fossil vegetation even from Europe seemed rather tropical in character. Nevertheless, there were problems: clearly death and suffering had been in the world long before there were people, for example. What was clear was that the inventive capacities of the Creator were not limited by the lack of rational spectators on Earth to applaud them. Millions of roses had blushed unseen; and conversely, there were now tracts, like the polar regions and the deserts, which were apparently void of life.

The danger was that bold thinkers might apply the ideas of uniformity and teleology to the history of the Earth, and see not Design but some law of progress. This was indeed what happened, first in France with Lamarck (*Philosophie Zoologique*, 1809);[8] and then in England in 1844 with the anonymous *Vestiges of the Natural History of Creation*.[9] This book, actually written by Robert Chambers but a well-kept secret, must have been a target for Whewell's anonymous publication, although the criticisms are veiled. Chambers believed the Sun to be a typical star, and that it had been formed along with the planets from whirling nebulous matter. Life had then appeared on Earth, and had gradually ascended through reptiles and mammals to man. The same process would occur elsewhere; and thus Chambers, like Huygens, would expect to find planets with rational inhabitants scattered all around the universe. They might not all have got as far as we have – Australia after all was zoologically behind Europe and America – but some might have got ahead of us, just as in time we could expect a higher species than us to emerge on Earth.

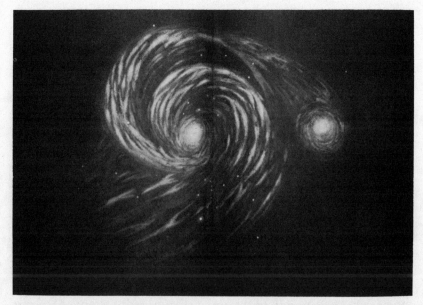

Figure 4   The great spiral nebula, revealed through Rosse's telescope, from J. P. Nichol, The Architecture of the Heavens, *London, 1850.*

Whewell was able to show that recent astronomical discoveries did not support the nebular hyopthesis – Lord Rosse's big (72-inch) telescope in Ireland resolved many nebulae into clusters of stars rather than suns and planets in the making. Whewell was not much bothered with whether or not there was some kind of progress through the strata revealed by geologists. For him, the appearance of man was the critical discontinuity in the history of the Earth. There was, for Whewell, 'no transition from man to animals'. Degraded and barbarous races of mankind were separated by a gulf from the most intelligent and valuable animals. This was an argument which befitted one whose academic career had seen a promotion from a scientific to an arts professorship at Cambridge, from Mineralogy to Morals. For him, if man were merely seen as a kind of animal, then the realities of responsibility and moral choice would be lost sight of. Plurality of worlds, with its emphasis on uniformity, went with ideas of evolutionary progress and thus with a materialistic vision of mankind. The objections to plurality of worlds were not so different from those to the speculative science of phrenology.

Whewell's hope was that science would not threaten human values if it avoided speculation and followed appropriate methods. He recognized, having read German philosophy, that the Baconian programme of inductive fact-collecting was never going to lead to real science, in which knowledge is organized. If sciences began with the right basic idea, then he believed that research would be guided (as it cannot be in a purely inductive system) along appropriate lines and science would steadily advance. His basic ideas differ from our 'paradigms' and 'research programmes' in that he believed that they would endure; philosophers of science in our century have learned to glory with Goethe in constant change. Whewell was generally uneasy about speculation, and believed particularly that the young should not be puffed up with such stuff: after a sound basic training in mathematics and classics one could then pass on to more inchoate studies. Such a notion lay behind much nineteenth-century advocacy of a liberal education: not only should there be only one culture, but young minds should be guarded against the dangers of conjecture. The Age of Science was a time when there was widespread agreement at least among the elite that a proper scepticism was in fact the best protection for both arts and sciences.

There was no ultimate safety even in scepticism – there can after all be no safety in the intellectual realm, or indeed generally, except perhaps in death – and it led in the mid-century to the theological work of Henry Mansel, and to the philosophy of Auguste Comte. Mansel's *Limits of Religious Thought* (1858) created a considerable stir and shocked some contemporaries who saw atheism lurking behind his apparent denial of much of Christianity. His insistence on the limits of the human mind was influential, and in his case scepticism had led to faith: he became Dean of St Paul's in 1868. But for many of a sceptical turn, like Frank Newman and A. J. Froude, religion seemed impossible to believe; and novels of religious doubt became a genre of Victorian fiction. Genuine faith in a person should be undoubting, as *Othello* shows us; unswerving belief in propositions on the other hand is not worth much, and it is a problem with religious faith to keep these two aspects apart.

Comte's scepticism took the form of positivism; his three stages of human thought, the theological, the metaphysical and the positive became famous. For him, science was positive knowledge, and must be kept free from accretions of obsolete thought. His scepticism thus extended into science: like Baconians he believed that theories should be no more than generalizations of observations, and that to attempt to go beyond the facts was hopeless. He became notorious for his remark that the chemistry of the stars was the sort of thing we could never study, just a few years before Kirchhoff started doing it with the spectroscope.[10] Indeed it is not infrequently the fate of the sceptic to be overtaken by events in this way. In France, Comte found himself inaugurating a 'positive' religion, and his followers became a sect; but his influence extended beyond them, and in Britain he found disciples particularly among religious doubters. This made his name and his philosophy suspect among churchmen; but his attitude towards theory also made him unacceptable to T. H. Huxley. Comte was perhaps not unlike Bacon, in that his writings on science recommended themselves in the main to those not actually involved in it.

To Cambridge men of science of the generation after Whewell's, the *Via Media* between scepticism and speculation seemed to be indicated by the wave theory of light.[11] Here, Young, Fresnel and their successors had adopted a model, the wave, and worked out how waves would behave in various situations. When tested for light, the predictions of the model always seemed to fit; and the criticisms of Brewster and others that the whole thing was hypothetical lost force with every success. George Stokes spoke for most Cambridge physicists when he declared that the task of a theory was not just to connect facts but give an account of the real operations of nature.[12] By the 1860s it could be said that whereas many chemists still seemed to doubt the real existence of atoms, physicists firmly believed in the ether and its light waves.

This was not as easy as one might suppose, for the ether was (as Lavoisier described phlogiston) a veritable Proteus,[13] with an extraordinary mix of properties, some apparently contradicting others: it had to be stiff and solid, and yet offer no resistance to the motion of planets or comets. Models of the ether involving

axles and idle-wheels were proposed, leading to the danger that
the models were more complicated than what they were devised
to explain. But at least difficulties were pushed further back,
which is one of the tasks of science; from light, one had got back
to ether. In a rather similar way, in the second half of the
century, the behaviour of gases was accounted for on the
assumption that they were composed of elastic molecules.[14] This
billiard-ball model, like the wave model, led to predictions which
always seemed to be verified; it looked as though in these models,
men of science had got hold of something like the basic ideas
hoped for by Whewell, on which firm and enduring science could
be built. There were sceptics, but they were seen as mere
metaphysicians.

The dynamical, or kinetic, theory of gases had not had such an
easy history as this might imply. It was originally proposed by
John Herapath in 1821 as a model based upon inelastic, because
absolutely hard, particles. Herapath was perhaps something of a
metaphysician, in that his hard atoms were the result of his belief
that ultimate particles cannot be springy because they can have
no sub-atoms to be pushed nearer together and bounce apart.
The incompressibility of atoms was not a new idea, but it did
mean that a theory of gases was much more difficult. Others
toyed with similar views, but it was not until the middle of the
century, when James Joule, Rudolf Clausius and James Clerk
Maxwell took it up, that it became a central part of physics.
They abandoned the hard atoms in favour of elastic molecules: a
model with hard particles gave the wrong answers, while one
with elastic particles worked well. Maxwell found that curious
predictions about pressure and viscosity were confirmed; and in
1859 he applied statistical theory so that instead of dealing with
one or two particles he could take into account the vast numbers
all moving at different speeds in different directions that make up
a real quantity of a real gas. In the end, not only could the kinetic
theory explain the gas laws of Boyle and Charles, but it could
also account for deviations from them. This is one of the great
triumphs of theory: it was similar to what had happened when
Newton's law of gravity had explained both Kepler's law of
elliptical orbits and its inaccuracy.

When the kinetic theory was found to make predictions in

accordance with those derived from chemical evidence – in predicting, for example, that oxygen was O_2 – it seemed as though its axioms must be true. The model must be a real portrayal of nature. It could then be used, at the beginning of the present century, to demonstrate that argon and helium were mon-atomic – A and He, rather than A_2 and He_2 – from data about their specific heats; their atomic weights could then be calculated, although at that time they seemed so inert as never to react chemically. Important chemical information had been inferred from physical evidence alone. There were sceptics, including William Thomson (Lord Kelvin), who was unhappy about the assumption of equipartition of energy in the theory; but success meant that even such eminent objectors were not heeded. It was familiar to physicists as to phrenologists that distinguished old men might not appreciate new ideas.

Thomson is famous for implying late in the nineteenth century that the main task of physics was done, and that what remained was to fill in the exact figures in equations already known. This was hardly characteristic of a man who at the celebrations of his fifty years as a professor at Glasgow could say that the word which characterized his efforts at understanding nature was 'failure'. More typical of him was to see at the turn of the century, in a lecture at the Royal Institution in April 1900, various 'clouds over the dynamical theory of heat and light'. The first cloud was concerned with the motion of the Earth through the ether; and the second with equipartition of energy. Two of the triumphs of nineteenth-century physics were starting to look a little uncertain; and indeed soon the ether theory was destroyed by Einstein.[15] The notion of wave/particle dualism, meaning that the wave theory covered only some aspects of light, together with the new atomic physics, brought serious modifications to the classic theories, which came again to be seen as models applicable under certain conditions but which could not be said to represent the way things are.

Thomson's scepticism was rather like that which has been forced upon scientists in the twentieth century. The provisional character of science has become evident, and like Davy we have had to accept and like sciences where much seems uncertain. Ideas of adding bricks to a building the foundations of which have been laid once and for all by Newton or Lavoisier no longer appear at all plausible. But in order to invest time, perhaps a lifetime, in science one has to believe that one is going somewhere; and Thomson, for example, was not really in despair about his efforts getting nowhere. There is an enormous amount that is firmly embedded and will probably survive any kind of intellectual revolution, and indeed which shows itself in technology. Thomson's kind of scepticism is really only needed in philosophical moments, or when asked to give lectures marking new centuries, or when things have got really stuck.

The scientist, in the nineteenth century and since, might thus admit that there may be more things in heaven and earth than he has dreamed of; but in his actual working moments he has to be a Horatio rather than a Hamlet because he has work to do. The nineteenth-century scientist therefore had little time for phrenology if he was working in physiology or anatomy, where there was plenty going on; and similarly because of his busy research programme Du Bois-Reymond could denounce Reichenbach's mysterious forces without bothering to repeat his experiments. There was simply no room in physiology for such things, though Du Bois-Reymond's materialism, which was not a necessary feature of the science of the time (for Kantian teleology was still a working tradition)[16] made the rejection more decided. Within a scientific field, there is an orthodoxy, and its limits are soon reached, except perhaps by the very eminent who cannot be stopped; and questions that cannot be answered are ignored. This scepticism is necessary, and goes with the fact that developed sciences cannot follow the Baconian programme of investigating anything and everything. What is interesting is that outside their own field scientists may be as credulous as the rest of us; scepticism is not a general requirement of the man of science. In the nineteenth century, there were thus eminent physicists and chemists who were seriously concerned with the spirits, for example, and others who patronized fringe medicine, such as homoeopathy. In our own day, Darwinism is denounced chiefly by outsiders (including some physical scientists) as dogmatism (which is odd from Fundamentalists), and defended from within by upholders sometimes unwilling to recognize that all science is provisional. Evolutionary theories right through the nineteenth century were contentious because they were momentous, casting light or darkness on man and his origin. It is to the study of animals that we shall now turn.

6

Debate about Animals

The first task of the natural historian had been to name. Even though insects did not answer to their names, they had to have them; otherwise nobody knew what anybody else was talking about. In part there was the problem of Babel – in different countries or different regions, plants and animals had quite different names; and in part it was a problem of taxonomy.[1] The distinctions between different kinds of creatures needed in practical life – perhaps between flowers and weeds, or the edible and inedible – may not be very useful across cultures, and may not be any good in science. Snails may be food or pests. When in the eighteenth century Mme Merian was preparing her sumptuous insect books, she found that there were no names in European languages for common but unshowy kinds of butterflies.[2] Nobody had been interested in distinguishing them. On the other hand, we should not underestimate our ancestors, or 'primitive' tribesmen of today: a study of the naming of birds in a part of New Guinea made recently demonstrates that their divisions and ours show an astonishing degree of similarity.[3] Out of 157 Kalam names, 96 applied to species, and 13 more to groups of species forming valid taxa in our system. There does seem to be something objective about the way mankind orders the world.

In the seventeenth century, people stopped writing in Latin even when writing for the learned, and various attempts were made at making up universal languages. An example was John Wilkins' 'real character' of 1668, in which symbols stood not for words but for things. Europeans had heard travellers' tales of how natives of different parts of China could not understand each other's speech, but could communicate by writing. Nearer home, our 'Arabic' numerals are examples of symbols of the kind Wilkins wanted: '2' may be read as 'two' or 'zwei', but to the English or German speaker it means the same. Yet artificial languages never have caught on. Wilkins and others in the early Royal Society were afraid of the metaphors associated with 'words' rather than things, and feared a science which was really word-play; but the invention and exploitation of models is akin to the exploration of metaphor, and science like everything else is a field for *homo ludens*. Only in mathematics do we see a successful artificial and international language, though when we come to talk about visual language we may see something a little like it; and music might count as a language, though one in which it is hard to say anything very definite.

In our natural history (as elsewhere later in science) it turned out best to adapt an existing language; but the man who did it, Linnaeus, in the middle of the eighteenth century, came from Sweden. His language was one spoken by very few foreigners, and indeed Latin had remained the learned tongue there as in Germany. Linnaeus devised a kind of Latin suitable for concise descriptions: animals and plants were not to be written up in flowery language, but in sentences containing many adjectives and few verbs.[4] His greatest idea came after he had written his first works and was doing an index for a book: this was to give a name composed of two words, a generic name and a specific or trivial one, to each species. This has survived for well over two hundred years and is still going strong as an efficient method of information-retrieval.

The difficulty is to know which are species. We, like the Kalam of New Guinea, distinguish jays and crows, or red deer and roes; but whether the two kinds of native oak in Britain are one species or two is a problem, and certainly we all know that there are more differences between different breeds of dogs than there are between some distinct species of animals. In the eighteenth century, as indeed for Aristotle, the best guide was fertility: a male and female belonged to the same species if they would mate and produce viable and fertile offspring.[5] This simple test showed that all mankind was one; but it was difficult to do with exotic or shy creatures. Birds in which the male and female looked very different were often classified as different species in nineteenth-century bird books; and the same happened with certain species of orchids which have flowers of different kinds, sometimes bafflingly on the same plant. With the rise of palaeontology about 1800 this test became anyway less relevant: we cannot supervise the mating of dinosaurs. Naturalists had to fall back on their judgement in family grouping; and some then and since have followed Adanson, a great French contemporary of Linnaeus, in trying to weigh up all characteristics of an organism instead of taking one or a few as crucial.

Naming good species is worth doing, but it is like grammar: a necessary step in the elegant and efficient ordering of ideas, but not in itself real science or literature. Just as astronomers sought for the pattern and the simple laws that must govern the movements of the stars and planets – making the world a real universe or cosmos, an ordered whole – so naturalists sought an arrangement which would make sense of all the different kinds of creatures that they found. For much of the eighteenth century, the old belief endured that everything was organized into a great chain of being, linking minerals, plants, animals and men (and perhaps angels). The explorer might hope to find 'missing links' in the chain living in remote places. This chain theory collapsed partly through its own incoherence under pressure – how do we know when we have got all the links? – and partly because of the discovery of past faunas. These might have been held to be missing links in a chain of which the links were very small in some places, and large (so far) in others; but this idea conflicted with the notion that God would have created the best of all possible worlds, which should therefore not change over time since all change must be for the worse.

Linnaeus' system of plants had a basis in arithmetic: he counted the male and female parts of the flower, and these figures placed it in its class. Usually this method put in nearby groups plants that intuition, or an enumeration as done by Adanson, indicated were alike; but not always, and if not it was just too bad. Linnaeus admitted that there was something artificial about this procedure, and hoped that in the course of ages a really natural system would be worked out. Such a method was proposed by Jussieu in 1789, and first in France and then elsewhere it was steadily adopted, and was improved as more characteristics were taken into account. But the difficulty about the natural method was that the classification had no shape.

Cuvier seemed to contemporaries to have had little trouble in demolishing the evolutionary scheme of his contemporary Lamarck.[6] Lamarck was eminent for his classification of invertebrates, and had in effect put the chain of being into motion, so that everything was moving slowly upwards, and creatures had not become extinct but had just changed, responding to their environment.[7] In his *Règne animal* (1816) Cuvier put forward a scheme in which the animals were divided into four great branches: instead of a chain we have a tree.[8] In classifying them he followed Aristotle in considering as many characters as possible; but he made great use of the 'principle of correlation', the way in which all the parts of an animal work together. Because of his interests in vertebrate palaeontology, bones were his especial interest. He believed that it would be possible to reconstruct an animal from a single bone – certainly, faced in the Paris basin with a valley of dry bones, he fired the imagination of contemporaries and of later generations by calling them back to life. Within his four branches, Cuvier subdivided using multiple criteria and separating, for example, the marsupials from the other mammals, although from the outside the koala looks a bit like a bear, and the thylacine (or Tasmanian wolf) very like a dog or wolf. His system became the standard one for half a century.

Animals could thus be placed on the different branches of the tree, but there was no symmetry about it all. One could not predict creatures not yet discovered, for the natural systems had no neat arithmetical basis like Linnaeus'; and anyone contemplating the system set out in Cuvier might feel (with King Alfonso of Castile on Ptolemaic astronomy) that if God had consulted him, he would have recommended something simpler. What did appear were various kinds of relationships, often called 'affinities'. This really means relationship by marriage, but naturalists used it to mean family relationship, recognizing that horses and donkeys were closely related even though they denied any actual evolutionary connections. Pre-Darwinian biology had a language which already possessed evolutionary overtones.

Another word often used was 'type'. This had a range of meanings, but was based on the idea that an individual could stand for a race, or prefigure other individuals not yet born. In this latter sense, preachers and theologians sought in the Old Testament (with its sometimes unedifying stories) types of the characters in the New Testament. In the interpretation of prophecy and of apocalypse this kind of reasoning was essential. In zoology, where again there seemed to be a former world of 'Dragons of the prime, That tare each other in their slime' in meaningless violence, the extinct animals could be seen as types of those now existing. Similarly, the good and great could be seen as noble types appearing before their time. Such thoughts went with an idea of progress, at least in the past; but such progress need not have been smooth, for the various faunas could easily have been destroyed and replaced in a series of catastrophes.

The other sense of type did not involve time. The first specimen to be formally described according to the rules laid down by Linnaeus (and later codified by Strickland and others from 1842 on) is the type, and these are carefully kept in great museums, which exchange duplicates after due comparison. If in doubt, one compares another specimen with the type in order to see if it belongs to the same species. In an extension of this, the first species described can be seen as the type, or the typical member, of its genus. Because Linnaeus worked in western Europe, it tends to be the European form which is seen as typical in this sense; and of course it is accidental which species in a group happens to be first described and hence seen as typical.

Types in this sense were used in teaching. The world is so full of a number of things that it would be hopeless to try to study every kind of animal or plant. The best plan was therefore to study a few typical ones: T. H. Huxley's course on the crayfish at South Kensington in the 1870s was based on this principle, close study of a crayfish preparing one for any invertebrate. Naturally the comparisons were closest with other crustaceans, and Huxley brought in shrimps, lobsters and crabs in the context of his course; the crayfish was a typical crustacean rather as the man on the Clapham omnibus was supposed to be a typical Englishman. But beyond this, Cuvier had demonstrated unity of plan and correlation of the parts in organisms. Once the crab had been shown by J. V. Thompson in his *Zoological Researches* (1828–34) to begin life as a little shrimp-like creature, and the barnacle perhaps more surprisingly to do the same before settling down in middle age, then the unity of plan in the group became more evident. Barnacles were a fit subject for Darwin to work on, as he did for many years, because of their curious metamorphosis; and also because of the way in which males have regressed to become merely vestigial in some species.

Given the whole family of crustaceans, it was possible to imagine an ideal or typical crustacean of which all the existing forms were more or less close realizations. This was done by Thomas Bell, a dentist and naturalist who achieved the feat of recognizing a shrimp previously undescribed in British waters in a dish of prawns served up to him in Bognor. No actual creature had all the limbs and segments of the ideal one; and their divergence from it was a measure of their height in the scale – crabs come above lobsters. Such a typical creature was a kind of Platonic ideal, imperfectly realized here below; and Richard Owen, Huxley's patron and then foe, made sense of zoology through a series of such 'archetypes', as he called them.[9] He saw homologies, unity of plan, everywhere and this gave him a pattern of a kind. The idea of types could help organize the science; and the exemplars of the type were often called 'representatives'. Thus *Alcedo atthis* is the only representative of the kingfisher family found in Britain. But the term could be extended, so that kangaroos might be said to represent deer in Australia because they live in something like the same way.

To most serious naturalists (and the term 'biologist', coined at the beginning of the century by Lamarck and Trew, took long to come into use), relationships which were not affinities were not really of much interest. But an exception was William Swainson, an outstandingly talented zoological illustrator, who took what we might call ideas typical or representative of his time further than most contemporaries were willing to do.[10] In Huxley and others the idea of types was a teaching aid, and Cuvier's branching tree gave some organization to his great tomes on the animal kingdom. Swainson had begun work on an encyclopedia of zoology, but this project was then transformed into a series of volumes for the *Cabinet Cyclopedia* organized by Dionysius

Lardner. Lardner was called 'the Tyrant' by his contributors; he was an Irishman, who created a sensation when he ran off with the wife of an army officer. He was one of those who saw the possibilities opened up by the cheaper paper, and the coming of machine-made cases instead of hand bookbindings, of about 1830. There was also a growing market for books as literacy increased, though there was not yet universal schooling: the Society for the Diffusion of Useful Knowledge, the Mechanics Institutes and the so-called March of Mind were features of the time leading up to the Reform Bill of 1832, and indicating a market for non-fiction.

Lardner's volumes were not quite an encyclopedia, since each one could stand on its own: volumes came out at intervals, and buyers do not seem to have been made to subscribe to the whole set. Swainson's original contract called for him to produce volumes at three-month intervals; he had already done preliminary work for his abortive encyclopedia, but even so the rate of production envisaged is astonishing, and it is not surprising that Swainson failed to keep up with his timetable. Faced with a whole series of books to write, Swainson had the choice of simply following Cuvier's system, which was coming out in an English version at the same time, or of doing something new; and he chose the latter.

Swainson was a man of parts, not altogether unlike Darwin in some respects. His career as a naturalist had begun when he was in the army at the end of the Napoleonic Wars; serving in Malta and Sicily, he met Rafinesque (later a founder of botany in the USA), who infected him with enthusiasm for natural history. After the war, Swainson went on an expedition to South America, to Brazil; but unlike Humboldt or Darwin, he brought little home, having been confined to the coast by a war. Back in London, deserting his native Liverpool which he compared unfavourably to Manchester, he was advised by William Leach at the British Museum to take up lithography. He duly produced the first work on natural history to use this new technique, *Zoological Illustrations* (6 volumes, 1820–33) and at the same time, *Exotic Conchology* (1821–2). The plates in these volumes were hand-coloured, and the original ones, coloured by Swainson and annotated by him so that colourists could get them right, have recently come to light and are at the Linnean Society, where his correspondence is also. He refers to the lithographed drawings as 'engravings', and he used many short lines like an engraver scratching on copper; pioneers are often conservative, and it was left to Edward Lear in his superb volume on parrots (1832) to exploit lithography, using bold flowing lines with his crayon on the stone.[11]

Swainson felt that an illustrator ranked below a real man of science, and he therefore also sought a reputation as a theorist. He adopted from an eminent entomologist, William MacLeay, the notion that the various classes of animals formed circular clusters (in which the extremes met) rather than diverging branches. These circles then touched others, forming groups of three, or (since the bottom circle could be made up of three little ones) five. There was only room in each circle for a set number of species. The group of circles formed a genus, and the genera could similarly be arranged in circles, and so on up the scale to higher groupings. Analogies played almost as important a part in the system as affinities did; animals with long snouts should occupy a place similar to birds with long bills, for example. The basis in threes was to be expected if God was a Trinity; and indeed in Swainson's high-church theology, we should expect to find in nature a network of analogies and symbols.

The great advantage of his system, with its reliance upon external characters and analogies, was that it enabled him to make sense of the whole animal kingdom without the lifetime of research which Lardner's schedule made impossible. He could not afford like Darwin to spend several years on just getting the barnacles straight. Knowing ahead of time what the classification of fish must look like is a great help if one is an expert only on birds and mollusca: and Swainson produced analogies which nobody had previously thought of. Another advantage in the 1830s was that there was no room among the apes for man. The circles were full up, and therefore man was not a part of the animal kingdom at all. Emphasis on analogies and external characters led to the splitting up of the marsupials, so that the thylacine was placed among the dogs despite its extraordinary reproductive system. Those who bought the *Cabinet Cyclopedia* therefore got a rather odd view of zoology, conservative in its

Figure 5   Orang-utan from the Penny Cyclopedia, *vol. 1, 1833.*

types and symbols but up to date in its appreciation of
J. V. Thompson's work, and rarely dull.

The circular or quinary system of Swainson interested some of
the more speculative minds of the next generation, including
Huxley, A. R. Wallace (who annotated his copy of Swainson on
the geographical distribution of animals), and Darwin (who had
copies made of MacLeay's writings in order to criticize them).
Swainson had got to know some of the leading zoologists in
London on his return from his voyage, and when writing the

scientific descriptions of the birds collected on Franklin's polar journeys he had gone to Paris and worked in the Museum there. But he had an immense capacity for rows, and fell out with everybody sooner or later. He tried to get a post at the British Museum, and was most indignant when it went to George Children, a friend of Davy and thus possessing superior patronage. Lacking a base in an institution, and having resigned from the Zoological Society after falling out with its Secretary, he was, as it were, writing his books in a void. A corrective to eccentric views, or a stimulus to close argument in presenting them, is provided by discussion or by teaching, formal or (particularly) informal; and this Swainson did not have. Even if one has good scientific ideas, it is never enough just to publish them: the world will never beat a pathway to the door of the scientific recluse; and all that the Avogadro or the Mendel can hope for is posthumous fame.

Swainson himself, in the first of his *Cabinet Cyclopedia* volumes, a *Preliminary Discourse* (1834) on natural history designed to complement John Herschel's on physical science, drew attention to the excessively formal meetings of most London scientific societies, where no questions or discussion followed a lecture or paper. The exception was the Geological Society, whose meeting room was even set out like the House of Commons for debate rather than in rows for a lecture.[12] Swainson also complained of the decline of science in Britain, and the lack of opportunities to study it. Other 'declinists' helped found the British Association for the Advancement of Science in 1831, meeting in different cities each year for lectures and discussions in various 'sections'; but Swainson played no part in it, thus avoiding the power struggles in which gentlemen of science usually with Cambridge connections came in effect to direct the organization.[13] Victorian science, perhaps Victorian intellectual life, had a certain robustness which could easily degenerate into quarrels, and it is a feature of the time that leading workers in the same field were often not on speaking terms: but at least in the BAAS good-tempered discussion was possible.

In the generation after Swainson's, one who was fully involved with scientific institutions was Edward Forbes.[14] He was also a man who got on well with almost all his associates, and his early death in 1854 was seen as a great blow to natural history. His career and his ideas form an interesting contrast with those of Swainson. He was a student at Edinburgh, where among other figures more in the mainstream of science and medicine he met Samuel Brown, a chemist famous (or notorious) for his trans- mutation experiments. He then made the pilgrimage to Paris in 1836/7, having renounced medicine as a career. In Paris he found that the lecturers did not as in Edinburgh confine themselves to one hour, but generally took an hour and a half or two hours; but he found his time there very valuable.

The problem about renouncing medicine in early-Victorian Britain was that there was no other very obvious way to live if one were a naturalist.[15] He tried giving freelance lectures in Edinburgh, and wrote a book on British starfishes (1841), but these brought in very little. A stream of scientific papers began to bring fame but not fortune. He hoped in the end to be chosen for the Chair of Natural History in Edinburgh, having a strong feeling (by then old-fashioned) that natural history was really one field. The problem was that there was no compulsory retiring age in the nineteenth century, and Robert Jameson worked on as Professor at Edinburgh for half a century, dying in office in his eightieth year in 1854. Forbes thus had to wait for fifteen years or so before the coveted post came up; during this time Jameson sometimes teetered but always recovered, and his courses and museum fell behind the times.

Forbes managed to get a place on a voyage in the eastern Mediterranean, where HMS *Beacon* was engaged in a survey. On his return, he got a post at King's College, London, teaching botany; but the fees were small. He became a keen supporter of the BAAS, founding a club which gathered each year at the meetings, for conviviality and science. They called themselves the Red Lions, after the name of the pub where they first met in Birmingham in 1839. Forbes' verses and caricatures became famous, and came to be typical of a kind of hearty humour forming a part of scientific meetings of the nineteenth century. In the end, a post was found for him in 1844 as palaeontologist with the Geological Survey, with a salary of £300 per year; and on the strength of this in 1848 he married. The Geological Survey was one of the very few organizations employing men of science in Britain – Greenwich Observatory was one of the others – but their status was about that of clerks in the Civil Service, and this rankled. At last in 1854 Forbes was duly appointed to the Edinburgh chair, but he died a few months later at the age of only thirty-nine; there was a general feeling that he had worn himself out with drudgery, and that success had come too late.

Forbes was thus a contemporary of Darwin and Huxley; but because he died at the height of his powers just five years before the *Origin of Species* was published, and because despite his time in Paris he remained an outdoor natural historian having little temptation towards laboratory-based physiology, his work was incomplete and soon seemed obsolete. In particular, his idea that the depths of the sea were lifeless was shown to be false with the deep-sea dredges begun in connection with the laying of telegraph cables beneath the oceans. It is to be expected that nature will surprise us; what was significant about Forbes' work was that he was a pioneer in studying the distribution of marine organisms, finding analogies between going deeper into the sea and going higher on land.

Forbes sought for general patterns in the distribution of living and extinct organisms, seeing a kind of polarity which governed the production of new types during past epochs. He loved to dwell also on the analogies between plants and animals, remarking in 1843 that 'both kingdoms seem pervaded by a double representation of each other'. He saw geology as disclosing the preparations 'for the reception of organized beings, a history which has all the character of a great epic, having for its hero, MAN'. This did not mean that he was at all sympathetic to evolutionary ideas, which seemed crude and mechanical; but he was not patient either with 'Bridgewater writing', because it meant seeing the hand of God more in some things than others. For him, every organism was an outward type of an inner Divine thought; so nothing was isolated, and everything was linked to everything else. He believed that the greatest scientists should also be poets, for both recognized affinity, analogy and beauty; and although to his biographer his mind seemed essentially Platonic, his greatest heroes were Aristotle, Linnaeus, Robert Brown and Goethe.

Swainson's and Forbes' energies were diffused in their own struggle for existence (though both hoped to live like gentlemen rather than just exist) and the need to write fast and to lecture for money and reputation.[16] Had Forbes lived longer, he might have founded at Edinburgh a school of naturalists as he hoped, and which was impossible at that date in London; and he might have written a major work. It was Darwin's fortune that spared him from the drudgery which wore out his contemporaries, and allowed him to do his research without worrying about whether he would have bread and cheese. Without private means, as a less-promising drop-out from Edinburgh medicine, he would even after his voyage have found it very difficult to find employment giving leisure for research.

One such job was at a museum. The British Museum was still one unit, containing both natural history and works of art, and there were posts there for a few naturalists. Robert Brown, the great botanist who had studied the distribution of Australian plants on and after a voyage with Matthew Flinders,[17] was the leading light there, with George Children as Keeper of Zoology; but the Museum never fulfilled the role of the Parisian natural history museum, with its lectures and research school.[18] When in the last quarter of the century, the natural history collections were moved to the new building in South Kensington, and had no longer to play second fiddle to the historical parts and the library, then the number of jobs increased and the scope for research improved. The first Supervisor (1856) of the whole natural history collections was Richard Owen, and he presided over the move to South Kensington (1881). Owen was an older contemporary of Forbes and Darwin, but lived far longer than either of them (1804–92); it is often a good idea to outlive one's opponents, but Owen's *bête noire* was Huxley (who outlived him), and by 1892 the world had become Darwinian and Owen seemed like a living fossil. He had moreover made himself disagreeable to many contemporary naturalists, and his posthumous reputation has been well below what it was in his earlier life, when he was seen as the British Cuvier.

Owen was a museum man from the beginning of his career.[19] After taking an Edinburgh medical degree, he was assistant curator of the Hunterian Collection of comparative anatomy at the Royal College of Surgeons from 1826. This collection had been formed in the second half of the eighteenth century by John Hunter who was largely responsible for raising surgery from a craft to a profession in England, and was the greatest anatomist and physiologist of his day. Its object was to show the comparative structure and functions of organs throughout the animal kingdom. With the coming of palaeontology, the museum was a great resource for anybody interested in reconstructing fossil creatures. It was also one of the attractions of London for the intellectually-minded tourist.

Cuvier had claimed to be able to reconstruct an animal from a single bone, but it was Owen who did it, working out that a bone from New Zealand, broken at the ends, must have come from an enormous flightless bird, and being vindicated when the complete skeleton of a moa was found later. This was indeed a triumph; like most of those who make predictions, Owen got some animals wrong later (identifying as mammals what turned out to be reptilian remains), but connoisseurship is a matter of probabilities rather than certainties. What really interested him, and what he tried to get across in lectures and in learned tomes, were the unities of plan, the homologies, to be found in the animal kingdom. This made him the intellectual heir of John Hunter, whose *Essays and Observations* he published with due filial piety in 1861. These were papers, filling two volumes, left unpublished at Hunter's death, which Owen dedicated to the Royal College of Surgeons, where from 1837 he had been Hunterian Professor.

Like Swainson and like Forbes, Owen saw the animal kingdom as exemplifying the ideas of its Creator, through a pattern of types and archetypes. He believed in a kind of progressive development of forms, but like Forbes stopped short of an actual evolutionary theory. He was one of the very few men of science who gave a welcome, albeit qualified, to *Vestiges* (1844) presumably because he could see the wood for the trees; but when the *Origin of Species* came out in 1859 he disliked it heartily, and became one of the most formidable opponents of Darwin and Huxley. Contemporaries saw some kind of jealousy here; Owen was not one who could brook a rival for the position of British Cuvier. He had (like Davy in the previous generation) risen by his professional competence to a position where he mingled with the eminent, like W. E. Gladstone, the Prince Consort and Samuel Wilberforce. Such success did not, however, always lead to social security. Huxley, the exponent of laboratory physiology, won a great victory over Owen when he contrived to get a statue of Darwin placed on the main staircase of the Natural History Museum; but justice has been done with the passage of time, and Owen's statue is now in pride of place in his building.

Rivalries are certainly a feature of intellectual life. In the case of Owen and Huxley there was a complex relationship, for Owen had been almost his father-in-science; but Owen had other serious reasons for seeing little in Darwinism. His Platonism meant that relationships were ideal patterns, not to be taken literally as lines of descent. His work on fossils and on comparative anatomy had convinced him of the reality of species and the lack of missing links between them. Moreover, he was a Christian, and saw Darwin's theory, with its apparent reliance on chance, as opposed to religion; materialism and chance were the bases after all of the Epicurean philosophy.

Owen was no fundamentalist, and many of his contemporaries (especially those working in museums, and classifying vertebrates) saw little for them in Darwinism: the family trees were hypothetical, and there was no independent evidence of real rather than ideal relationships. Those who were against evolutionary speculations were not 'creationists' proposing a different explanation of the origin of species, based upon one of the accounts of the creation in Genesis. High-church men such as Gladstone and Wilberforce did indeed stick closer to the literal text of the Bible than their twentieth-century successors would; they saw all its books as an inspired unit, 'Holy Scripture'. The 'higher criticism', examining the Bible as one would any other ancient text, only impinged on the Church of England in 1860 with the publication of *Essays and Reviews* by various authors including Frederick Temple and Benjamin Jowett. So before that most believers, like the young Darwin, took it for granted that everything in the Bible was true in a simple way.

Sophisticates might see it differently: thus J. H. Newman, writing in the *Tracts for the Times* (no. 85, 1840) when still a

member of the Church of England, drew attention to discrep-
ancies and differences of emphasis between various books of the
Bible. His intention was to show that the interpretation of
Scripture was too complex to be left to whoever happened to be
reading it. Only in the light of the traditions of the Church,
which had preserved these writings and formed a Canon of
orthodox books in deciding which writings should be included in
the Bible, could the meaning of the text be properly assigned.
Newman, with his concern about reserve in communicating
doctrine to those who were not ready for it, and his analysis of
assent into notional and real, seemed slippery to many contemporaries. To most Victorians, truth seemed a simple matter, only
confused by people like Pontius Pilate, of conformity to facts.
The Bible, then, must be true or false like a newspaper report;
few people had room for the truths of poetry or of fiction in their
religion.

Apart from this belief in the straightforward truth of books
which would now be seen as in differing degrees myth, with
Adam as Everyman rather than a historical figure, for example,
there was diffidence about the boundaries of science. The
Baconian suspicions of the speculative and deductive intellect
were strong; the public then as now liked science that was
hypothetical and interesting, and loved rows between experts,
but scientific opinion was cautious. And in particular, science
which was supposed to rely upon careful generalizing, with
repeated experiments and observations, could not handle unique
events. John Herschel in his *Preliminary Discourse* (1830) for
the *Cabinet Cyclopedia* urged that while astronomers can deal
with the orbits of the planets in the solar system, their inductions
cannot take them back to its creation. Thus one should avoid a
theory of the origin of species, which would inevitably be
unscientific, and concentrate on describing and classifying the
existing species. Darwin did not have to overthrow a coherent
and worked-out theory; he had to persuade contemporaries that
hypothesizing about origins was worthwhile.[20]

The famous example of a prominent scientist and Christian
who was something like a fundamentalist is Philip Gosse. His
Omphalos (1858) was written in the year before the *Origin of
Species*, so it is not a reply to it; but it can be seen as a reply in

anticipation. Gosse was perplexed at the vast tracts of time required by geologists to account for the deposition of all the strata. Since each stratum was accompanied by its fossils, it appeared that there had been animals upon the Earth for millions of years. To Gosse, this was just an appearance; and his book was 'an attempt to untie the geological knot', though it might seem rather to cut it. The omphalos was Adam's navel. Since he was not born in the ordinary way, Adam would not have needed one, but the general assumption was that as a man he had a navel. A bystander at his creation, or rather arriving a few moments after it, would feel justified in assuming that Adam had had a mother, been born and brought up in the usual way, and was in every way like us; but he would be wrong. For Gosse the Earth might be like Adam, supplied with what looks like evidence of its past history, but is not. The fossils are the equivalent of Adam's navel.

Gosse was a Plymouth Brother, and a marine biologist who had made aquaria popular and who ran seaside field-trip/holidays for those interested in natural history. His book did his reputation no good. It offended the religious, because it seemed to make God out to be untruthful or deceitful; in the scientific age, science was after all a route to truth, and only the Devil was the father of lies. Those who were seriously engaged in geology or biology could not allow such reasoning either, although in a sense it only represented the idea that science cannot deal with origins, carried to an extreme. Taken that way, the book can be said to have shown the problems of hypothesizing about origins; and also the problems associated with supposing that animals, men or plants were suddenly created, fully-grown so that they could manage on their own. Once looked at hard, Victorian 'creationism' could be seen to have its problems.

There were further difficulties when one began thinking about the numbers of animals required at the Creation. To sustain carnivores, there must have been many more herbivores; but the idea that two of each kind were created somewhere and slowly spread and multiplied would not go with real population dynamics. This was made explicit by Louis Agassiz, a Swiss who had in 1840 recognized marks of glaciation all over northern Europe and thus given us the idea of ice ages. He had also made himself the leading authority on fossil fish, taking over Cuvier's work. In 1846 he went to the USA to study and lecture, and in the following year was appointed to a chair at Harvard. He was one of the first eminent European scientists to make a career in the USA, and rapidly became a lion: his lectures and books were popular, and he built up a school and museum at Harvard.

American zoologists and botanists were, like those elsewhere, divided into 'lumpers' and 'splitters': lumpers are those who apply a coarse taxonomy, putting into one group individuals which may differ quite widely; whereas splitters use a fine taxonomy, making species where lumpers saw only races or varieties. Compared to Europeans, Americans tended to be splitters.[21] This was because they found that many of their animals and plants were very similar to European forms, but that they did differ in what seemed to be significant respects, such as size or colour. To Agassiz, who became the most eminent anti-Darwinian in the USA, these indicated that they were separate creations: animal populations had been created in about their present numbers and in their present locations. This curious conclusion set at nought the work on distribution done by Brown, Humboldt, Darwin and Wallace, and by the botanists Joseph Hooker of Kew and Asa Gray of Harvard; they had found all sorts of curious patterns, of which it seemed possible to make sense in terms of migrations, barriers and ice ages. Agassiz's kind of 'creationism' simply seemed to bar the door to further investigation of a question which could be scientifically handled.

It was also some way from orthodox Christianity, as 'creationism' (and indeed Bridgewater natural theology) can often or generally be. By 1857 when Agassiz's *Essay on Classification* appeared, as an introduction to a never-completed work on the natural history of the USA, it was already difficult to believe that Noah's flood had really been a world-wide catastrophe with animals surviving two by two; indeed Agassiz's work on ice ages had involved reinterpretation of data that seemed evidence of the Flood. His *Essay* was clearly incompatible with this story; and for man, the animal we are all most interested in, his account was very different from that in Genesis. By the nineteenth century, the idea that as children of Adam and

Eve we were all brothers and sisters under the skin had provided an argument against slavery. If black people had been separately created in Africa, and Red Indians in the Americas, and so on, then it might be as legitimate to enslave them as to enslave cattle. This was an important matter in the USA in the 1850s. To suggest that the races of mankind were separate creations was to depart widely from Biblical teaching. One could be a 'creationist' without being much support for Biblical religion.[22]

As a young man in Germany and in France Agassiz had met both an idealistic evolutionary system and the branching natural method of Cuvier and his disciples. His own system of classification involved a belief in the objective reality of both species and higher groupings – in fact, a kind of Platonism not so different from that of Forbes and Owen. Species should not, he believed, be defined in some operational way, based for instance on fertility. There were essential differences between them, so that arguments between splitters and lumpers were not about convenience, but about truth. Agassiz drew diagrams of the relative flourishing of different families of fish through past time, which showed gradual appearance and extinction – but he must have believed that species came into being and died out at definite times. In the absence of coelacanths, it was still possible to toy with the idea that species and genera had their life-spans as individuals do, even though some 'living fossils' among plants should have made this more difficult than it was.

By the 1850s attempts to understand the arrangement of animals had thus reached the limits possible if one assumes that species are fixed in time. Two-dimensional patterns like those of Linnaeus, Cuvier or Swainson made sense only against a background of idealism; by the 1850s they seemed to close doors to further enquiry rather than open them. Those working with invertebrates, such as Lamarck and Darwin, or with plants, such as Gray and Hooker, where the number of species was vastly greater than among vertebrates, were perhaps driven towards some more dynamic conception of a species, as something changing over time. The efforts of those like Gosse and Agassiz to make a definite theory out of the doctrine of Creation helped to show the value of an alternative approach; evolution might be made scientific.

In this way, Darwin could offer a 'working hypothesis', as Huxley put it, or a light shining in the darkness. In some of Darwin's own researches we can see how useful it was. He studied orchids, because of their extraordinary shapes; and argued (building on work of Brown) that they had evolved in conjunction with the insects which fertilized them. Instead of just describing and classifying the orchids, he could give some explanation of their appearance. This involved ecology, rather than just arranging dried specimens in interesting patterns. Quite new kinds of research were opened up, so that whereas the static kind of classifying was a rather sterile tradition, Darwinism presented an attractive research programme. Its trouble was that it could lead to just-so stories rather than to information; it was more provisional than some kinds of science.[23] But even the safest-seeming science is going to be obsolete sooner or later; obsolete taxonomy, like that of Swainson or Agassiz, is as dead as the dodo, and those who took the risks of Darwinism usually found it interesting and fruitful.

Darwinism never received a welcome in France, where in place of Darwin's historical and probabilistic method the way forward seemed to be in laboratory physiology. Here the models were from chemistry rather than from history; the idea was to give an account of how the parts of the organism were formed in embryology, and how they functioned in life. In a sense, this was a line which had been followed by John Hunter; but his heirs in Britain had not pursued it, partly because of a national aversion to the vivisection which seemed a necessary part of it and partly because it seemed materialistic. In Britain, Huxley introduced this kind of physiology from France and Germany from the 1850s; and in the last quarter of the century his pupils (trained in South Kensington, in what became Imperial College) filled chairs in British universities,[24] where the field was now becoming separated from medicine. In Huxley's own book *The Crayfish* (1879), written twenty years after the *Origin* had appeared, the development of crayfishes is a very unimportant part of the work, forming a confessedly hypothetical section at the end. Huxley as a popularizer of Darwin and as a teacher of biology emphasized rather different aspects of science. Huxley himself loathed vivisection, but encouraged it in his pupils and saw Britain's backwardness in the new physiology as something in need of urgent attention. Darwin, with his inherited money, doing little experiments in his garden and studying what we would call ecology, seemed rather like a survival from the eighteenth century: the new way to end arguments about animals seemed to be through chemistry. But that millenial day was still seen to be far off. Naturalists still needed illustrations, and it is to these that we shall turn next.

7

Discourse in Pictures

Natural history in the nineteenth century meant not only arguments about the system and man's place in it; it also meant great quantities of illustrated books and papers. And in technology too, the nineteenth century saw the coming to Britain from France of technical drawing, and the depiction of industrial landscapes and industrial processes. Scientific illustration meant not only pictures of animals, plants and rocks, but also of bridges, gas-works and railways. It is tempting to suppose that making pictures is almost devoid of theory; but no artist was ever really a kind of camera, and both theory and fashion entered into scientific illustration.[1] Just as scientific theory, from optics, geology, meteorology and phrenology, entered landscape and genre painting, so scientific illustrations show current outlooks. Artists' intentions are easier to recover in scientific illustration than in many other kinds of pictures. Since science is public knowledge, it is really printed pictures that will most interest us; and the nineteenth century saw the appearance of a series of new techniques which changed the appearance of scientific books radically.[2]

Books in the eighteenth century, and in the first quarter of the nineteenth, were little illustrated and were expensive. A handsomely produced but unillustrated book like Davy's *Six Discourses* (1827) would have cost more than a week's wages for an artisan like Faraday had been. Woodcuts were used for cheap publications, like the chapbooks sold by chapmen or pedlars; but these had crude illustrations. Woodcuts made on the side grain had become obsolete for scientific illustration by the middle of the seventeenth century. The splendid and spirited pictures in the early printed herbals had been done this way, but as increasingly detailed and accurate illustrations were demanded, plates etched or engraved on copper became the norm.

Woodcuts can be mounted with type, because what is to print black stands up from the surface; so it was easy to put pictures on the same page as descriptions, and herbals and bestiaries were liberally illustrated. Copper engraving or etching is not a relief process, but an intaglio one; what is to print black is cut down below the surface, and the copper is wiped before printing. A different press is needed, and a different paper is desirable, so to print pictures and text on the same page is awkward. Generally illustrations were separately bound in, with complex instructions to the binder as to where they should go – which he sometimes got wrong. Wicked persons are also tempted to remove the plates from their own or someone else's copy and frame them; the plates and text are much more separable in every way than they were in the old herbals. Sometimes the plates were even in a different format from the text, coming in a great atlas volume and making the work a nuisance to shelf.

Copper engraving was also a difficult technique. It was rare for the painter to be able to do his own engraving, and conversely few engravers were artists in their own right. Engraving is like translating; it has its own idiom, and skilled engravers became expert at transforming finished drawings or amateur sketches into plates, though engraving is a less rich language than drawing. The task was expensive: in the early nineteenth century, plates engraved for the Royal Society's *Philosophical Transactions* cost up to twenty guineas. These were plain plates, often very detailed anatomical studies in a large quarto format; if coloured, as some plates were in the Linnean Society's journal at the same time, the cost would be still higher.

For the Royal and Linnean Societies, both fashionable and scientific, money was no object and plates appropriate to the article were engraved. Authors were generally allowed to use the coppers if they later published their work in a book. At less exalted levels of society, economies had to be made; often many subjects were crammed on to a single plate. Authors had to economize in the number of illustrations they used, and readers had to get into the habit of turning a number of pages to find a

Figure 6   A pattern-plate of William Swainson from Zoological
Illustrations, *1821.*

relevant picture. Only at the upper end of the market could such
beautiful coloured works as *Plants of the Coast of Coromandel*
(1795–1820), published under the auspices of the East India
Company, or Audubon's *Birds of America* (1827–38), find a sale.
Audubon was an effective salesman for his book among the
wealthy, to whom he himself seemed an exotic. Colouring was
done by hand, following pattern plates coloured by the artist, so
that different copies of the same work, perhaps done several years
apart in a slow-selling book, may differ considerably among
themselves.

Just at the beginning of the century there came two new techniques, which reduced the price of illustration very considerably and brought visual language into the more ordinary works of science and natural history.[3] The first of these was wood engraving, done on the end-grain of boxwood with a burin rather than a knife. Thomas Bewick of Newcastle was the pioneer and perhaps the greatest exponent of this art; his *History of British Birds* of 1797–1804 was a splendid example of what could be done, although his text was of no scientific significance. Wood engravings can accompany type, and so once again pictures could go back on the page; they were also very durable, so that runs of hundreds of thousands were possible. The publishing explosion of the mid-century meant that scientific books were sprinkled with detailed illustrations done in this way. Copper in contrast was not durable enough for big editions, and where its effects were desired it was increasingly replaced by steel; but this was much less used than wood for works of science.

Wood engravings were generally small, because the box tree is small and its end-grain cannot accommodate a big picture. For magazines such as the *Illustrated London News* the solution was to bolt a number of small blocks together; but for full-page scientific pictures the technique of lithography became the standard one. Wood engraving was a technique which most artists could not manage, so a craftsman was needed to translate the drawing; on the other hand, an artist might be expected to do his picture on the stone, from which it would be directly printed. One attraction of the method was that an easy flowing line could be printed, in contrast to the short strokes made by the burin in copper; and Edward Lear's parrots, toucans and owls are stunning examples of lithography.[4] To get the detail which a copper engraving can show was more difficult, but by the middle of the century this had been achieved and lithography was the normal medium. Like copperplate, it needs a different press, and lithographs had to be bound in separately; but because the cost was about one-third that of copper engravings, there could be more of them. Only in about the last quarter of the century did colour printing, in the form of chromolithographs, become at all usual; and for expensive books, hand-colouring remained the norm well into the twentieth century.

Audubon was not the first to try to portray animals in motion, but his dramatic and vast illustrations of the birds of America were important as attempts to show living creatures. In fact they were painted from dead specimens, and to get within range of a naturalist was a misfortune for a nineteenth-century bird or animal; but Audubon painted them against realistic backgrounds, and in more or less plausible action. Some of his poses aroused derision from contemporaries, but his art suited the Romantic sensibilities of his day, and the principle of depicting living creatures became gradually accepted. In fact, this can hardly be done in many cases because the animals concerned are nocturnal or are very well camouflaged; but the stiffly mounted corpse on its studio stump, so characteristic of the eighteenth century, soon came to look dated. Those engaged in classifying creatures in museums worked anyway with dead ones; it was only at the very end of the nineteenth century that the collecting of dead animals and dead birds gave way to the careful observation of living ones, using binoculars. The naturalist in the field might not even have seen his specimens alive, but have acquired corpses from local people.

So there was often an element of delusion about zoological illustration. This is increased if the picture is systematic, that is, if it is to show members of close species. Such illustrations are valuable if one is discriminating specimens, in a museum or perhaps in the field; but close species are not usually seen together, because they will fill slightly different ecological niches. A systematic plate can also easily date, as classifications change: what seemed species are reclassed as races, and what seemed close species are put into different genera. In such a plate, visual language is being used to put across a theoretical message; and like ordinary language it can become obsolete. To the historian, this makes it more interesting because it presents hostages to fortune, and there is more to interpret.

Another kind of plate with more than one animal or bird on it is the ecological one. Audubon had often shown a predator with its prey, and the relatively static bird plates of John Gould's artists for his splendid volumes also show this.[5] Edward Lear, drawing parrots and toucans in the zoo, got them very lifelike but left his backgrounds very sketchy, letting the bird stand out brilliantly but giving very little clue about its habitat. Pictures done in the zoo can be much better than those done in a museum from a dry skin, perhaps not very convincingly stuffed; but only if there are field sketches too for the artist to work from can he hope to get the background right. Only rarely did the artist actually go to where he might see exotic species in the wild[6] – and even if he had gone there always, he would probably have found them very hard to see. But at least he could, if armed with field sketches, show an animal on a food plant, perhaps; or depict various species which might be found in the same rock-pool or stretch of veldt.

We can see this process happening with different plates of the Australian Diamond Firetail finch. This was shown in a handsome but stiff pose, painted by Barraband, in Vieillot's book *Tropical Songbirds* (1805–9):[7] though drawn from life, it looks stuffed but is very decorative, and is accurate enough to fit its primarily taxonomic purpose as well. A few years later, in Lewin's *Birds of New South Wales* (1813), we find a pair displayed, giving us more information, by one who had seen them wild; but the background is rather vague local colour. This book was the first natural history book known to have been published in Australia, where the engravings were also done – probably by a transported forger. Then a generation later, in Gould's *Birds of Australia* (1848) a plate shows a pair on grasses which form part of their diet. Since this was lithographed, by H. C. Richter, it is also much bigger, in a large folio size instead of quarto; but with small birds which can be shown life-size on a page of ordinary dimensions, this merely adds splendour.

There is little theory-loading in any of these plates, and none of them attempt to catch the bird in motion as the 'field guides' of the twentieth century do. The lie of every feather is indicated, and while the plates give great aesthetic pleasure, their object was to help identify birds hit by a collector. If the specimen first collected has perished, the picture may indeed be the type. Sometimes, as when a plate of a fossil shows its matrix in detail, or that of a fish shows the river where it was caught, we are given more information than we needed – though such details may help to balance the tone of the plate and thus have an aesthetic function. Later illustrations brought in more: they included the whole life-cycle of an insect, as the plates of butterflies and moths of Humphrys and Westwood (1841, 1843–5) do; in their handsome but crowded pictures we find a number of close species, on characteristic food plants, thus getting both system and ecology. It is curious that in modern works on insects, the different stages are often shown in different illustrations, in order to be clearer, to give less systematic hostages to fortune, and to use colour only where it is necessary.

By the time Darwin was working on the *Origin of Species*, the man who was to become his great opponent in the USA, Louis Agassiz, was using diagrams to show how the various families of fish had flourished and declined through the geological epochs; and in his handsome volumes of 1857, *Contributions to the Natural History of the United States*, I and II, he included a plate of turtles to show the range of variation within a species. All the three turtles shown belong to one species, but differ considerably; it was upon the existence of such differences that Darwin was to base his theory of natural selection leading to divergence and eventually to the emergence of new species. Similar diagrams to show the periods of flourishing of various natural groups were used by Richard Owen, Darwin's chief adversary in England; illustrations which look Darwinian may not be so in fact, but merely indicate that contemporaries were all wrestling with similar problems, and were sharing a good many assumptions. Owen took over from cartographers the convention that where outlines are unknown one should just dot them in; so his pictures of fossil bones show conjectural restoration of broken or lost pieces. It was perhaps unfortunate that in verbal controversy with T. H. Huxley he was less careful.[8]

By the 1870s Darwinians were using diagrams showing hypothetical family trees to account for relationships among species. Specifically Darwinian illustrations are to be found in R. D. Fitzgerald's *Australian Orchids* (1882–94), one of the most splendid books published in Australia in the nineteenth century. He showed in two cases the flower with the insect which pollinated it, following Darwin's lead in his *Fertilisation of Orchids* (1862). There Darwin had argued that the extraordinary shapes of orchids were the result of evolution in conjunction with generally a single kind of insect fertilizer. Fitzgerald's plates were thus tactfully laden with theory, and contrast with other flower pictures which show decorative insects, or insect plates which show food plants.

What is striking about natural history illustrations is sometimes their longevity.[9] Thus bird books today sometimes use pictures from John Gould, well over a century old, or from C. G. Finch-Davies (*Birds of Southern Africa*, ed. A. Kemp, 1982) or G. E. Lodge (*Unpublished Bird Paintings*, ed. C. A. Fleming, 1983), dating from the early years of our century. Despite all the changes in theory which have happened in the interim, these plates still apparently have value, and pass time's test. Sometimes the name originally given is no longer apt, because of revision of genera or species; and sometimes close species put on one plate, as with Lodge's pictures (which are of New Zealand birds), are now separated; but these things do not seem to matter too much to the publishers and editors. In the past, such longevity has sometimes been a sign of static science. Thus a picture of a chimpanzee from Edward Tyson's *Orang-Outang* of 1699 was used in the standard English version of Cuvier's *Animal Kingdom* (1827–35), and sections through different woods from Nehemiah Grew's *Anatomy of Plants* (1682) were reproduced in Davy's *Agricultural Chemistry* (1813). In both these cases the old plates made the necessary points, and there was no need to go to the trouble and expense of getting new ones. If there has been little theoretical change, or if they are works of art in their own right – and especially if they are decorative – then illustrations can have a useful life much longer than the text which accompanied them. In most fields the nineteenth century was the age of the textbook, such as Thomas Thomson's in chemistry and Lyell's in geology, where successive editions made the earlier ones obsolete; the plates that survive usefully are perhaps to be compared to the very few classic books like the *Origin of Species* which go on selling.

In the days of the Bestiary and the Emblem Book, animal pictures had shown the world of Aesop's fables; butterflies giving us a symbol of resurrection, and also of feckless enjoyment, in pictures waiting to be decoded. By the nineteenth century this language of animals and flowers had been almost lost, though E. H. Aitkin in India in 1883, drawing a chameleon hunting a butterfly, could whimsically call it 'the pursuit of pleasure'. At the beginning of the century George Montagu on the title-page of his *Testacea Britannica* (1803) made a verbal and pictorial point about the Chain of Being; but the text, a verse from Stillingfleet, was essential. Strictly visual language to make scientific points had been used as early as 1555 by Pierre Belon in his *Oiseaux*, where he put the skeleton of a man and a bird side by side to demonstrate their unity of plan, or homology. Works on comparative anatomy continued this tradition, though by the nineteenth century there was more caution about using mankind. As we saw, Montagu's contemporary James Parkinson, the radical surgeon, made a visual point in the frontispiece to his *Organic Remains of a Former World* (1804–11) which shows Noah's Ark grounded on a rock in the middle distance with the rainbow beyond it, while on the beach in the foreground are various shells of extinct creatures which had missed the boat.[10]

Parkinson knew nothing of dinosaurs, revealed in the work of William Buckland, of Oxford,[11] and of Gideon Mantell, a surgeon of Lewes, in the 1820s. Cuvier had reconstructed extinct mammals, but here the analogies with living elephants, bears and rhinoceroses were close enough for nobody to doubt what they must have looked like. To encounter a hairy elephant would be alarming; but the dinosaurs caught and have held the imagination of us all because they seemed dragons. Apocalyptic visions of monsters seemed to be realized in these terrible eyes and terrible jaws. The palaeontologist was an Ezekiel calling to life the dry bones; but in this case the old sign 'here be dragons' was apt, and the prophet was bringing diabolical creatures to life – perhaps a Frankenstein rather than an Ezekiel.

Since they were safely extinct, nobody needed to have real nightmares about dinosaurs; but their reconstructions did bring problems, and involved theory-loading.[12] Ichthyosaurs, with their fishy shape, were no great problem; but Mantell's dinosaur was called Iguanodon because of some resemblances to the Iguana. It was imagined that like lizards it went on all fours; bones were mounted this way in museums, and at the Great Exhibition of 1851 hollow model dinosaurs were made to entertain the public. Inside the Iguanodon men of science were entertained to dinner; this was possible because the beast had been made horizontal. Dining at ease inside a dragon makes a good symbol of the harnessing of mighty powers of nature for man's comfort, or perhaps of pride coming before a fall. But soon Owen's erstwhile disciple T. H. Huxley began to question the dinosaurs' relationships to lizards; he saw the closeness of birds to some reptiles, and inferred that at least some dinosaurs had gone on two legs rather than four, the Iguanodon among them. How one reconstructed a dinosaur might depend upon one's views of relationships between reptiles and birds; and indeed with the coming of cladistics, taxonomy based on descent, this is still a live issue, though there is now general agreement that Iguanodon stood on its hind legs.

Dinosaurs as dragons rather than as ancestral birds or lizards struck the painter John Martin as he prepared a frontispiece for Mantell's *Wonders of Geology* (1838).[13] Martin was a specialist in vast canvases showing the destruction of Sodom and Gomorrah, Belshazzar's Feast with the Writing on the Wall, and such apocalyptic scenes; and his frontispiece is suitably dramatic, though small. In the foreground three dinosaurs, clearly wingless dragons rather than definite species, are fighting in mud or slime, watched by a pterodactyl with its wings spread like a cormorant. Further away two more monsters threaten each other, while in front some tortoises and ammonites show the inhuman scale of the dinosaurs. Curious palm trees dot the scene, and the background is immense and desolate, as in Martin's paintings. The 'country of the Iguanodon restored' is very different from Victorian Sussex; the strange light which picks out the struggling animals also reveals how many tons of animal life the artist believed that the primeval world could support on quite a small area – as later dinosaur pictures always tend to do. The chief effect is to highlight the horror of the desolate and pre-human world of purposeless suffering. If man was a late creation, his Fall could not have brought pain and death into the world; ferocious dinosaurs, and animal suffering generally, brought grave problems to those seeking a theodicy, justifying the ways of God.

Grey Egerton, describing in 1852 a fossil fish which he had lithographed for the Geological Survey, could take a more cheerful view, calling it a 'pretty little specimen'. This, like the

Figure 7 'The Country of the Iguanadon', frontispiece to Mantell's Wonders of Geology, eighth edition, 1864.

advertising language which Darwin's family teased him with using about beetles and barnacles, was characteristic of Victorian naturalists. Darwin delighted in the exquisite adaptations which he saw in living creatures, as Paley and other natural theologians had done before him: contemplating fossils (or nature generally) could bring pleasure or pain depending on one's temperament and state. The glum and the cheery would at least have agreed that getting the best artist for one's pictures was crucial, and some rows, as for instance between Albert Günther and Francis Day, were over attempts to commandeer an artist. In the case of Günther and Day, who were eminent ichthyologists, the artist was G. H. Ford. Fish are very demanding for the draughtsman because one cannot put them into an attractive or anthropo-morphic pose; what can sometimes be done is to show the scenery where the fish was caught. A view of the Tweed behind a salmon gives local colour; and more seriously it can serve, like the blue wash behind a bird, to get the tonal balance of the picture right.

To get the specimen from overseas to the artist was the normal practice, but sometimes it was possible to take the artist, or to send him, to the animals. Thus on Captain Cook's first voyage, Joseph Banks the naturalist took a number of artists, including Sydney Parkinson who was to specialize in natural history subjects.[14] He produced many excellent pictures before his death on the voyage; and later expeditions benefited from professional and amateur draughtsmen. Graphic records from nineteenth-century voyages of discovery are very splendid.[15] But sometimes a naturalist or collector overseas would find and perhaps train local artists to depict what he collected. This happened in China, and some of the attractive paintings collected by John Reeves in the early nineteenth century have recently been published at the British Museum (Natural History). Many of these were stock pictures rather than specially commissioned ones; Chinese artists were skilled at rendering plants, birds and animals, and could work in conventions attractive to European merchants.

It was not so easy in Latin America,[16] where J. C. Mutis led a Royal Botanical Expedition in the late eighteenth century; this lasted into the early nineteenth century, for he died in Colombia in 1808. Mutis found Indians to paint flowers for him, setting the plants stiffly down the middle of the page in a way that looks exotic to the European eye. Some of the plants they drew were diseased; it is always a matter of difficulty to know which features of a plant or animal are individual peculiarities and which are specific characters. The artist, especially if he has only one specimen to work from and cannot therefore generalize, must paint what he sees; the greatest of artists in this line will, however, agree with Josef Wolf that 'we see distinctly only what we know thoroughly', a maxim he adapted from Goethe. Mutis' volumes began to come out only in 1954, which must be some kind of record; in this case the explanation was the various political vicissitudes of the century and a half following his death.

William Daniell in his *Animated Nature* of 1807–12 had tried to convey not only background but also atmosphere. But it was Wolf who succeeded best in the nineteenth century in bringing drama into natural history illustration.[17] In his plates for John Gould's bird books the character of the bird comes through, so that the birds of prey are not just accurate but menacing. In his *Life and Habits of Wild Animals*, Wolf drew a crocodile seizing a tiger before the eyes of its mate, a gazelle dislodging a leopard from its back by dashing under a low branch, and such exciting episodes – the animal equivalents of the genre paintings in which Victorians delighted. Wolf's animals look right even in these scenes where the temptation to make them anthropomorphic is great, because the picture might be taken to have a moral for human beings. These animated plates may take theoretical risks: crocodiles may turn out to dislike the taste of tiger; territorial displays may be misinterpreted as mating ritual; a fight may really be a kind of dance. In W. L. Buller's *Birds of New Zealand* (1888) the eminent artist J. G. Keulemans showed a Kea attacking sheep.[18] This was a time of panic, in which one farmer alleged that a thousand of his sheep had been killed by these parrots; but it is now known that they rarely kill sheep. An alarming picture encapsulated a false belief. On the other hand, careful observation of living creatures had led to the solution of some problems, such as how the toucan tucks its head under its wing to sleep – depicted in E. T. Bennett's book on the London Zoo (1831–2).

Wolf's exciting plates were engraved by J. W. and Edward

Whymper, the famous mountaineer; but by Wolf's day the new technique of photography was beginning to come into use in scientific illustration. One may feel that while an artist may fudge a picture, the camera cannot lie. On the other hand, a photograph cannot be generalized or simplified to bring out the important diagnostic features of a species; it must be always of an individual at a particular time and place. There is still scope therefore for the artist, but no longer for the run-of-the-mill artist, in scientific illustration.

The first photographs required long exposures, and animals cannot be expected to sit still as long as Victorian adults did; but by the last quarter of the nineteenth century fast exposures were possible. Darwin's *Expression of Emotions* (1872) had photographs of humans, including a splendidly sulky child; but the animals were drawn, because the 'disappointed and sulky' chimpanzee, or the dog 'in a humble and affectionate frame of mind', would have been harder to photograph with the equipment available. In the same decade, Edweard Muybridge developed high-speed photography of men and animals in motion, and was at last able to clear up the question of how horses gallop. Even such curious things as the trajectory of a crow's wing were photographed before the end of the century, while photography through telescopes and then microscopes proved invaluable in astronomy and in biology. Where photographs can show things invisible to the unaided eye, they are extremely valuable; and in the second half of the nineteenth century the camera, a gift of science to society, could be a useful scientific instrument. As well as recording movements, it preserves for us portraits of scientists, and pictures of industrial processes.

Natural history illustrations often included dissections, but it is striking to find in J. S. Miller's *Crinoidea* (1821) an exploded drawing of one of these 'stone lilies' or stalked starfish. With an engineer's interest in how all these tiny 'bones' had fitted together, Miller had gone far beyond his contemporaries dealing with less complex fossils, who drew in with dotted lines parts missing in their specimen. At this date such illustrations were uncommon even in engineering, where drawing was becoming both a serious matter and something like an art form. It is

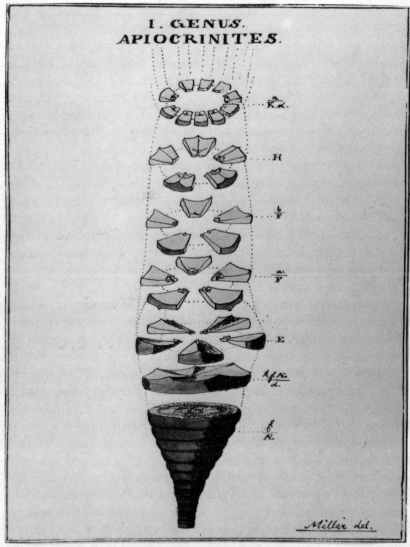

Figure 8   *Exploded drawing of a stone lily from J. S. Miller,*
Crinoidea, *1821.*

curious that although in the nineteenth century the historical
picture was the most prestigious, and the anecdotal picture the
most popular, it is the scientific illustration – birds, flowers,
ships, steam-engines, and great buildings such as railway
stations – which catch and hold our attention and admiration.

The engineering drawing was an essential feature of the industrial revolution as it moved into the stage of large-scale works; it is thus a characteristic element of the scientific age of the nineteenth century.[19] Its origins go back to the rediscovery of perspective in the Renaissance, and then to the architect's drawings of the eighteenth century. The earliest steam-pumps of Thomas Newcomen and others in the eighteenth century were not erected from the kind of detailed and accurate drawings customary by the nineteenth. There was much use of natural materials, and craftsmen expected like medieval masons to be given general rather than highly specific directions. In the same way, shipwrights were expected to show initiative when building in the traditional way in wood; but with the coming of iron and then steel as materials, leading to bigger ships as steam replaced sail, calculations led to exact drawings, which became the vehicle of centralized control and stronger industrial discipline.

In the last decades of the eighteenth century, James Watt prepared careful working drawings for the steam-engines which Boulton and Watt were making. These engines required much better fit of components than had Newcomen's; parts bought in had to fit exactly, and then those putting up the engine and its house on site required to know just what to do. A language of technical drawing began. Watt's were usually coloured, the colours at this period sometimes indicating materials, sometimes function, and sometimes being just decorative.

Contemporary with Watt was Gaspard Monge, who in 1795 published his *Géométrie descriptive*; although this book was not directly influential in Britain for some time (British engineers having little sympathy for abstract mathematics), it formed the basis for modern engineering drawing. Monge provided a system for clearly displaying three-dimensional structures in two dimensions. Careful shading, as was done by Swainson drawing shells and by James Sowerby and his family drawing minerals, could indicate exact shapes; and projections from different angles showed how things fitted together. By the middle of the century, an engineering firm would have a large drawing office (which by the end of the century might include women draughtsmen) in which detailed plans indicating exactly where rivets must go, and more highly finished drawings for the client, were prepared. The draughtsmen were by this time becoming separated from the engineers, in that specialization which was such a feature of all aspects of life in the age of science. The engineering drawing itself accelerated this process, because projects carefully worked out in advance on paper could be carried out by many workers each doing a small repetitive part of the whole operation, perhaps in distant factories and workshops. The skilled man could be replaced by the factory hand.

It was a feature of the history of technology in the nineteenth century that an invention by a craftsman, or a traditional practice, was opened up to scientific analysis; once properly understood, it could be improved upon. Some theoretical science, such as thermodynamics, depended upon this process. Sadi Carnot came to the second law of thermodynamics from pondering on the efficiency of an abstract and idealized engine;[20] and Helmholtz came to the first from considering simple mechanical devices such as trip-hammers driven by water-wheels. Renaissance drawings of machines, and even some of those from the eighteenth century, sometimes display an ignorance of how the device must really have worked; and at other times they are fudged, so that important features cannot be seen because the artist did not understand them. A proper engineering drawing cannot be thus fudged; like Wolf, the draughtsman must fully comprehend what he is drawing. As a concise and precise international language like that longed for by the natural philosophers of the seventeenth century, where symbols would stand for things and not words, engineering drawing was valuable in opening machinery to scientific study.

This had indeed been one of the aims of the scientific societies of the seventeenth century, such as the Royal Society; and it was best realized in the superb plates in the *Encyclopèdie* of Diderot and d'Alembert which show French manufactures in the middle of the eighteenth century, before the industrial revolution changed techniques.[21] At all times, but increasingly in the later eighteenth and the nineteenth centuries, new techniques have been brought in by observing their use elsewhere – what would now be unkindly called industrial espionage – and sometimes by tempting away workmen. Apprentices trained by the great engineers such as Robert Stephenson introduced the best practice in their subsequent careers; they were therefore sought after, and a high premium was charged when a boy was taken on. Apprentices were trained to make drawings rapidly, so that when they saw some machinery with which they were unacquainted they could sketch its salient features; this would correspond to the field sketch of the naturalist. Engineers and manufacturers of the nineteenth century had to keep an eye open for the rather-too-well-informed casual visitor, making notes and sketches of what was going on. The early industrial development of the USA largely depended on this kind of borrowing. A sketch by an expert can convey much more information than several pages of prose, as we all know from the instructions that accompany household apparatus.

At the beginning of the nineteenth century, there was almost no standardization in science or in technology. Drawings were not done to a standard scale; the metric system was just being introduced in France and its conquered territories, where it replaced a range of different miles, fathoms, feet and inches; and there were no standard interchangeable parts, even nuts and bolts and screws being individual. It was not until the 1870s that in British science grams and centimetres replaced ounces, troy or avoirdupois, and cubic inches; while in engineering the old units lasted well into the twentieth century. Variations of units having the same name meant that sometimes there was easy money to be made, buying big chauldrons and selling small ones; but usually this was only a minor nuisance, entailing conversion tables and making international communication a little more difficult.

In Britain at the beginning of the century the naval dockyards were among the largest industrial concerns; and there Marc Brunel first introduced high technology in the form of machinery for making pulley-blocks for the rigging of warships. His son, Isambard Kingdom Brunel, is famous for his railway work and for his ships. The principle of interchangeable parts did not catch on in British industry as fast as it did in the American gun industry; Colt's revolvers were the great examples of what became known as the American system of manufacture.[22] In the later nineteenth century Europeans found themselves borrowing techniques from the USA. This was particularly true in the military field, as the old smooth-bore muskets were replaced by rifles at the time of the Crimean War of 1854–6.

Because of the interest Colt's guns aroused at the Great Exhibition of 1851 in London, commissioners were sent to the New York Industrial Exhibition of 1854. One of them was Joseph Whitworth, who prepared an interesting report on some of the manufactures and on the patent system of the USA. He is particularly important in our story because of his work on standardizing the pitches of screws and of nuts and bolts. Faraday had written in the 1820s that it was a good idea to have the same pitch on various bits of apparatus, so that they could be connected together in different ways; but this idea had not become general. It was an important step towards the ideal of interchangeable parts, and it went well with the nineteenth-century transforming of craft activity into modern industry, remotely controlled by paper in the form of plans and drawings, prepared by people in a distant office and perhaps in a distant town.

Great Exhibitions were an important visual argument for science and industry, and an opportunity for patriotism; and it is to the diffusion and dissemination of science that we shall now turn. Science was not merely loved or used, as understanding or as power; it was also believed in, and therefore it was also preached. To audiences ranging from the Royal Family to working men, the message of science was put across by earnest evangelists and ardent publicists. Since scientists had to eat, much of this rhetoric was also concerned with the need for the age of science to support its science with money and honour.

8

Scientific Culture

Outside London, the major agency for the diffusion of scientific knowledge and the spread of the gospel of science was the British Association.[1] At its first meeting, this had been essentially a group of provincial enthusiasts; but they were soon brought within the fold of the established Church Scientific, the 'gentlemen of science' mostly from Cambridge, who took upon themselves the direction of the efforts of the provincial amateurs. The first meeting had been at York in 1831; thereafter the Association visited the university cities of Oxford, Cambridge, Edinburgh and Dublin, before in 1836 venturing to Bristol and subsequently to other large towns. The visit of the Association was not unlike the coming of a circus: local intellectuals had an opportunity to mingle with the eminent who came in droves, cities competed in the lavishness of their entertainment, and in days before radio and television one could see and hear well-known speakers addressing scientific audiences, or sometimes great crowds of working men. In the scientific sessions one might if one were lucky witness some great row.[2] Meetings were often held in museums or educational institutions specially built or refurbished to be ready for the great occasion.[3]

We can see what went on by looking at the volumes of the annual *Report* of the Association; for instance that of 1855, containing the proceedings of the meeting at Liverpool in September 1854. It is a stout book, giving the rules of the Association, the officers at all the meetings so far, and all the life and annual subscribers, with addresses – making it a kind of 'Who's Who'. Holding office in the Association was an important indicator of status in the scientific community by this time. There were by 1854 seven sections, each with its own President, Vice-Presidents and Secretaries: Mathematics & Physics; Chemistry; Geology; Zoology & Botany; Geography & Ethnology; Statistics; and Mechanical Science, which meant in effect Engineering.

Each evening there were General Meetings, in the Philharmonic Hall or St George's Hall: at one the President, Lord Harrowby, gave an address; at another, Owen talked about the anthropoid apes; and one evening Edward Sabine talked about terrestrial magnetism. On two evenings there were demonstration-experiments, in electricity, optics, magnetism and gyroscopes, with an exhibition of photographs of the Moon; and on the last evening was a meeting where the proceedings of the General Committee and the allocation of grants were explained to members. During the days, the various sections met to conduct their own meetings, hearing papers from both eminent and unknown men of science. The meetings of the Association helped both to raise the consciousness of scientists and also to enforce some specialization: members expected to attend only one of the sections.

The President began the meeting as usual, with a review of the scientific advances of the previous year, naming experts who had helped him in making his list. Perhaps the development of most interest to us in 1854 was that on an American initiative there had been an international conference at Brussels on shortening ship's passages by producing better charts of winds and currents. This led in Britain to the setting up the Meteorological Office under Captain FitzRoy, formerly of HMS *Beagle*, who went on to make the first weather forecasts in Britain.[4] Sarcasm and theoretical criticism of these in due course preyed on FitzRoy's mind, and in 1865 he killed himself; after an intermission, and despite 'scientific' criticism of the empiricism of it all, the weather forecasts were resumed, perhaps with slowly increasing accuracy down to our own day.

Harrowby was an eminent politician, who had been for many years MP for Liverpool; his scientific interests were in geography and statistics. He hoped, in his address, that Finance and Political Economy might be removed from the sphere of party contention; that the spirit of science might be imported into politics. He was horrified at the 'monstrous forms of misery and vice' disclosed by the statistics, believing that facts must disturb prejudices and theories, and that the Census reports were 'the Doomsday-book of the *people* of England, as the great volume of the Conqueror was of its *surface*'. He called the attention of his audience to the question of a decimal coinage – realized nearly 120 years later – and to the excitements of living in an age of gas and steam, of railroads, telegraphs and photographs, in which 'the importance of science is no longer questioned. It is a truism – a commonplace.'

Nevertheless, he found it worth enquiring, as Presidents of the BAAS were wont to do, whether 'sufficient facilities for education in science exist or are in progress in our country; and whether Government or other important bodies provide sufficient encouragement and reward for its prosecution'. Until recently he believed that 'the magnificent State of Britain' had done little for the Science upon which her wealth, power and very existence depended; except to provide security and freedom in a country its citizens could be proud of, and this had indeed been enough. But spontaneous vigour of citizens and of political and religious institutions was not, happily, any longer felt to be sufficient. Governments were now 'heartily and honestly engaged in repairing the deficiencies of centuries', while institutions were reforming themselves, so that soon without the excessive Continental systematization, there would be excellent practical arrangements for scientific education. The Government was even prepared to put money into it.

What about rewards for scientists? Harrowby believed (wrongly!) that titles were not what scientific men would want: their names anyway 'were a passport into any society, the proudest of the land'. Public life and science are both jealous mistresses, not tolerating known devotion to another pursuit; but at least those who administer the affairs of the country ought to, and do not, 'know enough of science to appreciate its value, and to be acquainted with its wants and with its bearings on the interests of society'. This should soon be the case with all educated men; and might be brought about sooner with formal contacts between the BAAS and Parliament. He had no fears that the pursuit of science need crush the imagination, or lead to a tedious style; or that it would ever become the favourite study of the many, overgrowing and smothering ethical and elevating influences, though perhaps too passive, from Literature and Art.

After this exercise in Victorian Values by a maverick Tory, we pass to various commissioned reports on the state of different sciences. There is one on the facts of earthquakes, arranged in tabular form and number three of a series – a contribution to 'Humboldtian' science, occupying 326 pages.[5] This is followed by a report on the construction of lifeboats, then by one on the state of knowledge on radiant heat, again the third of a series, essentially reviewing recent work all over Europe; then an account of the results of magnetic observations at observatories in the British colonies, another Humboldtian exercise; a report on how Solar Radiations affect the Vital Powers of plants, grown under bell-glasses of various colours; a catalogue of meteors; a report on the surface of the Moon; another on water-pressure machinery; and reports on the equivalence of starch and sugar in food, on deviations of the compass in iron and wooden ships, and on the vitality of seeds. All these were themes of some contemporary importance, and, as we see from the Treasurer's reports, money was voted for the expenses of the various eminent people drawing them up. The Report indeed includes a statement of all the grants made since 1834: over £1,500 in 1839, 1840 and 1843, but later falling to £159 in 1849; at this meeting £751 was voted.

For the scientific sections, there are merely notes and abstracts of papers, which nevertheless occupy 176 pages; they are separately paginated and indexed from the formal reports, which makes these publications confusing to cite. Some 'abstracts' are long; one, in small type, on the climate of Nova Scotia occupies over 11 pages; others are so brief as to be merely the title, such as 'On the lost tribes of Israel', under Ethnology, and 'On the decomposition of Magnesian Limestone at Brodsworth', under Chemistry. The subjects covered include the costs of a recent strike at Preston; the phenomena of diamagnetism; the heating effect of secondary electric currents (much electricity was still in Chemistry at the BAAS); the extent of palaeozoic glaciers; and the action of literary and philosophical pursuits in promoting longevity. There was in this vast menu something to please everyone.

With its rituals, its entertainments, its clashes and its uplift, its patronage and its exposure of the eminent to provincial crowds, the BAAS came to occupy an important place in the intellectual calendar of Victorian Britain. Railways and steamships brought delegates from all over the country to the chosen centre each year, and also brought foreign visitors, giving an international flavour to the occasion. Distinguished foreigners were appointed Corresponding Members; there were fifty-five in 1854. The newspapers described what went on, particularly the annual Presidential Addresses, which generally called for more scientific education, and were fully and widely reported. It is curious nevertheless how poor the primary sources sometimes are for debates at BAAS meetings: the famous clash between Wilberforce and Huxley in 1860 at Oxford, in the new museum there, has scarcely more contemporary accounts than that between St George and the dragon, so that it is impossible to know exactly what happened.[6]

The BAAS was based upon a German model, and in its turn formed a model for associations in other countries, such as France and the USA. But because of the industrial importance of nineteenth-century Britain, and because of its provincial vigour, the BAAS was particularly prominent.[7] The themes raised in Harrowby's address were indeed important ones for his audience. There were few in Britain who combined scientific and public life: Leonard Horner, John Lubbock and Lyon Playfair might be examples. Here there is a contrast especially with France. The first Napoleon had been a patron of men of science,[8] perhaps hoping that finance and political economy might be thus 'removed from the sphere of party'; and men like Berthollet, Laplace, Cuvier, Gay-Lussac, Arago, Dumas and Berthelot played prominent parts in, or received titles from, governments of different colours during the century. In the USA, the Coast Survey, the Topographical Engineers and the Smithsonian Museum received government support, playing a part in the promotion of trade and the winning of the West.[9] In Germany, universities were controlled by governments, and also supported by them: the British ideal of self-help was not widely shared.

More serious was the lack of careers in science even by the 1850s.[10] Edward Forbes, T. H. Huxley and John Tyndall all had

long and unhappy waits for suitable posts; and even insiders
from the ancient universities could not depend upon posts if they
had scruples about being ordained. In 1847 Richard Sheepshanks
wrote of John Couch Adams, a Cambridge mathematician who
had just predicted the new planet Neptune: 'I think there is a
hope that Mr Adams will continue his astronomical career. In
any other country there could be no doubt of it, but in England
there is no certain *carrière* for men of science. The Law or the
Church seizes on all talent which is not independently rich or
careless about wealth.' Sheepshanks, a leading figure in the Royal
Astronomical Society, himself fitted into all three of his
categories: he was both a lawyer and a clergyman, and did not
practise either profession because he had inherited wealth from
his cloth-manufacturing father. Adams was in a more difficult
position. He was offered a knighthood in 1847 when Queen
Victoria visited Cambridge, but he declined it and the honour
which came to him was accompanied by little emolument.
Because he did not wish to be ordained, his Fellowship at
St John's College, Cambridge, lapsed in 1852; but he was
fortunate in getting another at Pembroke College, and then in
1858 was elected to the Lowndean Professorship of Astronomy
and Geometry at Cambridge, which did not require that the
holder should be a clergyman. For contemporaries the moral that
they order these things better in France was particularly easy to
draw, for Neptune had been independently and at the same time
predicted by U. J. J. Le Verrier. He not only entered politics,
but had a chair of astronomy created for him, and was in
1854 appointed Director of the Paris Observatory. In this case a
nasty priority-dispute between the 'seconds' of the two men
ended in mutual respect, with Adams as President of the Royal
Astronomical Society later presenting Le Verrier with its gold
medal.

The BAAS was on the surface a democratic and provincial
body, run in fact by a small group of professionals, a Cambridge
network of gentlemen of science.[11] In astronomy, this system
worked well because there were series of observations which
amateurs could make which really advanced the science; of
comets, meteors and red stars, for example. In natural history the
same was true, and the careful observer and recorder was a very

important person in the advance of the science.[12] Here both foot-soldiers and officers were required, and indeed they still are: what churches call stewardship of talents has a place. But in chemistry, physics and physiology, the laboratory sciences becoming prominent in the nineteenth century, there was little scope for the amateur. There were a few people who had private laboratories, and who were like Lavoisier amateurs of fully professional competence; and there were those who lived by consultancy. But the advance of the laboratory sciences depended essentially on stewardship of money: that is, on financial support. At the beginning of the century, chemistry and physics had still been fairly cheap: the expensive sciences were astronomy and exploration with its associated natural history, the Humboldtian programme; but by the latter half of the century physics and chemistry, though still to our eyes heavily dependent on sealing wax and string, were becoming the big science of the day.

At the beginning of the century, the Royal Institution had been founded with two aims which turned out to be incompatible: to bring science both to mechanics and to the landed gentry.[13] The mechanics lost out, and the place became a research institute funded by subscribers who attended lectures. Davy and then Faraday turned out to be brilliant lecturers, holding large audiences for courses on scientific topics, each lecture being illustrated by experiments, often of a spectacular kind: Davy made a model volcano filled with potassium, and poured water into it; and Faraday flung the fire irons and then the coal-scuttle to a large electromagnet overhead. With Tyndall the tradition continued, and indeed it still does.

These lectures making science comprehensible and palatable for the fashionable, the earnest and the intellectual circles of London were not the whole programme of the Institution. Some courses appropriate to medical students were given; but there were also lectures which had no connection with science in our narrow modern sense. Thus the famous wit Sidney Smith lectured on ethics, and S. T. Coleridge on Shakespeare. Recent reprints of Royal Institution discourses, organized under disciplines, do not really give the flavour of the programme as a whole. By the middle of the century the most important lectures were called Discourses, and were delivered on Friday evenings. Often they

Figure 9 *Scientific lecture theatre from C. F. Partington,* Natural and Experimental Philosophy, *1828, vol. II, frontispiece.*

were by a member of the staff of the RI, but the majority were outside speakers. The event was fixed up long ahead, and in the correspondence of the eminent delivering such a Discourse appears comparable in importance to giving a major address at a BAAS meeting. From 1851 they were published as *Proceedings of the Royal Institution*, sometimes in abstract but generally in full. They were printed fairly rapidly: Francis Galton gave a lecture on the nature and nurture of men of science on 27 February 1874, and sent out offprints on 9 April, at a postage charge of ½d.

By the latter part of the nineteenth century there were two aspects of the popularization of science. The first was to put it across to laymen, interested in science but inexpert in it. Royal Institution audiences included the eminent in various walks of life, and the lecturer had an opportuntiy for evangelism among the mighty. From the beginning the audiences also included women, though no women held forth there in the Age of Science of the nineteenth century – when the Curies came to England in 1903 to talk about *le radium*, it was Pierre and not Marie who gave the Discourse. Women could not play a conspicuous part in scientific life. The second aspect of popularization was informing active scientists in other disciplines about what was going on. This kind of higher popularization was an aspect of the specialization which was such a feature of the nineteenth century, and since. At the BAAS the different sections reinforced these disciplinary frontiers, sometimes drawn a little differently from today's but on the whole amazingly stable: the intellectual map of nineteenth-century Europe is more like today's than the political one is.

Of the scientific Discourses at the Royal Institution some were describing original research, occasionally very new as when J. J. Thomson reported in 1897 the discovery of the electron (if we may allow that phrase); others were in the nature of a review of recent work. At the beginning of the century, all journals except that of the Linnean Society were general journals,[14] publishing material in all (or most) of the sciences – even the Linnean Society included animals, vegetables and minerals within its purview. While not every article would interest every reader, or be comprehensible to him or her, they were usually written in an expansive style so as to be accessible. The Royal Society at the time when the BAAS was founded was a club having a majority of non-scientific members, headed by a Royal Duke.[15] Just as the word 'scientist' was being coined in the 1830s to express a new feeling of community, specialized societies and journals were coming into being for astronomers, entomologists, chemists and other groups. Aiming at a very specific readership, they carried shorter articles written in the jargon of the various separate sciences. The books of Mary Somerville (e.g., *On the Connexion of the Physical Sciences*, 1834) proved valuable to men of science wanting to keep up with what was going on in other fields, and unable to understand it all even if they had had time to read it.[16]

Hers was high-level popularization; at other levels were such elementary works as Jane Marcet's *Conversations on Chemistry*, (1806) and Samuel Parkes' *Chemical Catechism*, (1806) though both these reckoned to be up to date with new discoveries such as those of Davy. The style of the letter, the conversation, the catechism, or the lecture appealed to those popularizing science and afraid that the neophyte might be put off for life by too dry a style. As articles in scientific journals became more formal, the need for popular writings and lectures increased. When we look at the lectures delivered at the Royal Institution, we find a striking variation in the level at which topics are treated, and an amazing range of topics; what is uniform is the skill with which the lecturers on the whole wrote persuasive prose. At the Royal Institution, one could get popular science of a high standard; and what was offered gives us an idea of what was seen as important.

By the 1870s things had changed much since Harrowby's BAAS address of 1854. There was universal elementary education in Britain[17] – a century behind Prussia, and three behind Saxony. Science was being taught in public schools and in grammar schools, and there were courses in universities, which in Britain began at about this time to catch on – this was the era when the great municipal universities started to flourish, though it was to be some time before they received government grants. What the lectures at the Royal Institution seem to indicate is that despite these changes the age of 'two cultures' was still some way off; at the elite level, those interested in the advance of science also wanted to hear about other cultural activity. Science and scientific argument were still widely defined.

Another way of putting this would be to say that scientific culture was remarkably wide. The Royal Institution was a centre of science, and the idea was that Discourses there would be scientific; when literature or religion or painting was under discussion, it would be from some scientific point of view. Just as Harrowby had hoped to bring politics within the orbit of science, so the lectures at this focal point in the metropolis indicate how science permeated all Victorian culture. Those who see the appearance of humanities in the programmes of scientific institutes in London or the provinces as an intrusion might sometimes in fact be seeing a take-over bid. Men of science in the nineteenth century were increasingly becoming specialists; but the eminent were often interested in other things as well, and hoped that these were not irrelevant to their science, or that its methods were foreign to them. Some saw literature and art as a relaxation from science, but others saw it as an extension.

General lectures were important for a scientific community becoming increasingly fragmented. The Royal Society had by the 1870s become an academy; its elections were an accolade, bestowed after agonizing by a committee. It had government grants to dispense, and scientific leadership to undertake. Its Presidents now served for short terms, and were always very distinguished men of science. It no longer functioned as a club; members could not understand what all the others were up to, and were aware that the interests of physicists and of biologists might be opposed. The implications of scientific advances, and the value of new pieces of science outside the narrow group of professionals, could be spelt out at the Royal Institution; it was a centre for evangelism, and for a meeting of minds trained in various traditions.

The sixth and seventh volumes of the *Proceedings* of the RI cover the years 1870–75. Some of the Discourses printed in them are no more than a title, but most are published in full. Some are curiosities: on 8 March 1872 R. Lieberich, of St Thomas's Hospital, lectured on Turner's paintings with reference to certain faults of vision. He remarked: 'It is a generally received opinion that Turner adopted a peculiar manner, that he exaggerated it more and more, and that his last works are the result of a deranged intellect.' The lecturer was convinced of the incorrectness or injustice of this opinion; in his view the explanation of Turner's beginning in middle age to paint what we think of as 'Turners' was that the crystalline lens of his eyes became rather dim when he was about fifty-five, 'and dispersed the light more strongly, and in consequence threw a bluish mist over illuminated objects'. He compared Turner's last paintings with Beethoven's last works, written while he was deaf: in both cases, some people preferred them.

In all the arts, we may prefer the abnormal and regard that which is entirely sound and healthy as commonplace; but late Turners have gone too far, in Lieberich's opinion, to be admirable. He found that Turner by 1833 was painting trees unknown to any botanist; probably he painted them because he saw them that way; and indeed with the help of a lens the lecturer could turn a picture of a common tree into a Turner tree for the audience to see for themselves. Lieberich went on to describe the functioning of the eye, and to diagnose astigmatism in a Parisian portrait painter; seen through a suitable lens, his productions looked much better. Colours also may get distorted as the eye lens gets yellower with age; in some pictures of Mulready he believed he could demonstrate this defect.

There are philosophical problems about this kind of approach to paintings: if the artist paints so that what is on the canvas looks to him just like what is in front of him, then the astigmatism or other defects ought to cancel out and his picture look all right to us. More to the point is that the Discourse indicates the scientism of the period: it is taken for granted by the lecturer that Turner ought to paint a tree of a recognizable species, for example, and assumed that portrait painters are after an exact likeness. Faraday was a friend of Turner and an admirer of his work, and might well have been more sympathetic to his attempts to catch the effects of light in paint. In 1873 Walter Pater's book, *The Renaissance*, first appeared; its Epicurean message that art is to help us maintain ecstasy is far from the view of Lieberich, in the light of whose lecture we can see why the aesthetic movement of the late nineteenth century came about – in reaction to representational theories, in the age of photography.

Lieberich is a representative figure in an important group in the intellectual, scientific and industrial life of Victorian Britain. This was a group of Germans who, like Prince Albert, had come to seek their fortune, after getting an excellent education in their homeland. There was indeed a steady brain drain; and some, like Lieberich, had intended to come for a visit rather than to stay permanently. Others in this group were Max Müller, the pioneer of linguistics, who lectured in June 1870 and again in March 1871 on anthropological topics; and William Siemens, the engineer, who gave Discourses in March 1872 and May 1874. We may note that Lieberich also spoke again at the RI, this time on the real and the ideal in portraiture, in March 1875; by then he had both an MD and MRCS after his name. The Siemens family maintained both British and German connections, for William's brother Werner remained in Berlin and built up a great electrical engineering business there, making among other things the first trams.

The Prussians had been allies of Britain at Waterloo, and pre-industrial, Biedermeier, Germany was very attractive to many Britons. The universities there seemed models for British ones, and gradually the feeling gained ground that Germany was the intellectual centre of Europe – in science, in philosophy, in classical studies, in theology and in history Germans were making the running. Prussia was the weakest of the victorious powers in 1815, and posed no threat to Britain, as France seemed to do when in the 1860s Napoleon III appeared to be following the ambitions of Napoleon I, and contemplating invasion. When in 1870 Prussia went to war with France the general feeling was that the French were the aggressors and that Bismarck might perhaps teach them a lesson. The rapid victories of the Germans astonished their friends; in part they acted (like Sputnik later) as an immense stimulus to scientific education, because the German victories were attributed to their science and their efficient schools. They also led to some strategic thinking: it was clear that Prussia was now the dominant land power in Europe, and that the defences of Britain were by no means impressive – much of the army was always away in the colonies, and the navy on distant stations.

This is reflected in the RI Discourses. In 1870, on 18 March, the civil engineer and FRS J. F. Bateman lectured 'on the subway to France'. He urged the building of a channel tunnel, at a cost of eight million pounds, with a pneumatic railway running through it. This would be the first stage of an overland communication with India; and it would also be of immense advantage for Britain's trade with Europe, valuable to all parties. In July 1870 the war began, and suddenly the Channel became a first line of defence rather than an obstacle to trade. In 1871 there were three Discourses having some connection with military topics, whereas in 1870 there had been one; in 1872 there was one, and in 1873 none again. The lectures in 1871 were from E. J. Reed (chief constructor of the navy) on 10 February, on the stability of ironclads following the loss of HMS *Captain*; then on 3 March Captain Noble, FRS, late of the Royal Artillery, reported on experiments on the power of gunpowder, some of them carried out with William Armstrong, the armaments manufacturer; and on 12 May Colonel Drummond Jervois of the Royal Engineers spoke on the defence policy of Great Britain.

Having spoken of the vital importance of the navy for a trading power, Jervois turned to the urgent need to protect London from an invader, because London was not like Moscow, which the Russians had abandoned before Napoleon and survived, but like Paris. 'The fall of London would render further resistance impossible', because the seat of government, the heart of empire, the centre of commerce, the focus of communications, and the only government manufacturing arsenal, would all be in enemy hands. One answer was to construct three or four fortified entrenched camps about twenty or thirty miles from London as indirect protection, manned with regular troops; another, to build about fifty works, about twelve miles from London, which being purely defensive could draw upon the vast resources of the city in men and materials. To invest such works a besieging army would have to be at least 700,000 strong, and London would be able to hold out a very long time, as the example of Paris showed; the Prussian siege there lasted five months, and had things gone a little differently Paris might have been relieved. The navy in the nineteenth century may have been an insurance policy for free trade, but Pax Britannica was not something which could be taken for granted, even by Victorian Britons.

Ever since the days when Lavoisier and then Davy had worked

on gunpowder there had been close connections between science and defence; but the quantity of hard science in these papers is small, as it is also in F. A. Abel's on substitutes for gunpowder (17 May 1872) where he spoke of nitro-glycerine and dynamite and gun-cotton, and even in John Tyndall's of 4 June 1875 on Whitworth's rifled guns. What these Discourses show is that 'science' still meant something wider than it does in the twentieth century; what the lecturers were encouraging was realistic assessment and sound judgement. Some lectures were some way even from this loose definition of science: John Ruskin lectured on Verona and Italian art in 1870, and Richard Westmacott, who was both a Royal Academician and FRS, on art five weeks later – but he talked particularly on the question of assembling and conserving the Greek sculpture in the British Museum. In 1873, Edward Dannreuther talked on the music of the future (meaning Wagner's) and Sidney Colvin on taste and artistic judgement; and in succeeding years there were always some Discourses essentially on the arts – a word which by this time had come to mean literature, music, painting and sculpture rather than crafts and techniques.

The Royal Institution had just moved its laboratories up from the basement where they were no longer valuable in 'our struggle for existence', as William Spottiswoode put it in January 1873. The laboratory of Davy and Faraday had been like a large kitchen of the early nineteenth century.[18] Such laboratories were not meant to be public places, and the requirements of furnaces and sand-baths and enormous wet-batteries all pointed to the basement as the best place, although without good artificial light the hours of usefulness of basement laboratories was limited. By the 1870s heating was done by Bunsen burners rather than furnaces; and a laboratory which at the beginning of the century had been as good as any was now 'a mere makeshift' or a 'noble relic'. There were by then 'splendid edifices' in London, Oxford, Cambridge, Manchester and Glasgow, and to retain researchers at the RI would have been impossible without new accommodation for their research. Competition in facilities at home and abroad would prevail, Spottiswoode believed, for a generation to come – we know that it has gone on ever since. The era of sealing wax and string was beginning to come to its end, although the end was far off; in the Cavendish Laboratory under Rutherford in the 1930s the story was that research workers asking for string were made to say how many inches of it they needed.

Most of the Dialogues are about the kind of research carried on in the new laboratories which were becoming a feature of life by the 1870s. Tyndall, Professor of Natural Philosophy, talked in 1870 on one of his special interests: dust and disease. He, like other Darwinian evolutionists, was strongly opposed to the idea of the spontaneous generation of life from inorganic matter, though no doubt it must have happened once at least in the past. He repeated and extended Pasteur's work on micro-organisms, 'our invisible friends and foes', and played an important part in getting the germ theory of disease accepted in Britain; and this led to an interest in the particles suspended in the atmosphere – especially that of Victorian London, but also in more salubrious places. In 1871 he again talked on dust and smoke, describing a respirator he had invented using charcoal to absorb noxious fumes; this device to assist firemen was in the Royal Institution tradition going back to Davy's miner's lamp.

Tyndall also lectured in these years on radiant heat, and on acoustics. His other great interest was in climbing, especially in the Alps; a subsidiary peak on the Matterhorn is named after him, and the Zermatt museum possesses his ice-axe and his signature (and those of other Victorian intellectuals) in old visitors' books. In his books *The Glaciers of the Alps* (1860), and *Hours of Exercise in the Alps* (1871), we find him describing daredevil climbs, astonishingly ill-prepared by modern standards; but also making a serious study of how glaciers move, which in 1872 came out more formally as *The Forms of Water*. Tyndall did not in these years hold forth on glaciers at the RI, but on 13 May 1870 Canon Henry Moseley (father of an eminent zoologist, and grandfather of an eminent physicist) gave a Discourse on the descent of glaciers. Unlike Moseley, Tyndall was far from orthodox in religion; he was an agnostic, but his alpine books show a kind of religious attitude to a universe he believed impersonal; he was a keen Darwinian.

By the 1870s Darwin's evolutionary ideas had come to prevail generally, and also to act as a paradigm in fields remote from his own.[19] Thus on 14 May 1875, John Evans, FRS, spoke on the 'coinage of the Ancient Britons and Natural Selection', showing how copies made from Roman models gradually degenerated towards simple and symmetrical patterns, saving trouble for the makers. Two weeks later Colonel Lane-Fox (later Pitt-Rivers), a pioneer of scientific archaeology, spoke of 'the evolution of culture': some kind of Darwinism entered into all kinds of explanations, as tends to happen with highly effective theories which get extended far beyond their sphere and come to function almost as metaphor.

More directly Darwinian themes are to be found in other Discourses, such as that of Burdon Sanderson (5 June 1874) on insectivorous plants. Darwin had himself studied some of these plants, verifying that they do actually consume the flies which land in their pitchers or on their leaves and get caught there, and finding that it was apparently to nitrogenous compounds that the plants responded. G. J. Allman lectured in March 1873 on another Darwinian theme, the formation of coral islands. Before he had ever seen one, in the early days on HMS *Beagle*, Darwin worked out that fringing reefs, around islands in an ocean whose bed was slowly sinking, might maintain themselves at sea-level. When the island had sunk beneath the waves the former fringe would be left as an atoll. The crucial point in this deductive piece of science was that corals grow only in shallow water; the theory became generally accepted, and helped to make Darwin's name as a geologist.

In April 1873 W. H. Flower, subsequently to be in charge of the British Museum, Natural History, in South Kensington,[20] lectured on palaeontology and the support it gives to evolutionary theory. In 1859, when the *Origin of Species* had come out, the evidence from fossils was suggestive but very incomplete: geologists had sought to characterize strata by their fossils rather than to follow family trees through from early times to the present.[12] Here as elsewhere, Darwin's theory showed its value as a working hypothesis; and by the 1870s Flower could argue that all the hoofed animals' general line of descent and relationships could be worked out using fossil discoveries. The history of the horse, worked out from American fossils, was merely the most spectacular. Looking for lines of descent with modification had not merely led to finding interesting fossils, but had also produced better understanding of how existing species fit together.

The voyage of HMS *Beagle* has come to eclipse those of the series of scientific voyages to which it belonged, because of the eminence of its naturalist-passenger. Only Cook's voyages, the first in the series, are anything like as famous: even the polar voyages of Parry, Franklin and Ross are relatively little known, and in Britain the French, Russian, Spanish[22] and American scientific expeditions in the same tradition are forgotten. With such voyages, it is hard to decide exactly how 'scientific' each one was; but certainly mapping, terrestrial magnetism, gravitational measurements, botany and zoology, meteorology, anthropology and the study of ocean currents were a part of all of them. By the 1870s general questions of marine biology had become prominent, partly in an evolutionary context in which life had first begun in the sea. Edward Forbes had suggested that the bottoms of the oceans must be lifeless, an azoic region, because light never penetrated there. This hypothesis gave later biologists something to test, and in the event to prove false, thus indicating the truth of Bacon's dictum that error is better then confusion, and of Popper's idea that science works through conjectures and refutations.

It was to be expected that Britain as the leading maritime power should take a lead in scientific voyages, which could be directly linked to shorter passage times and greater safety, and the general advancement of trade. The great promoter of such voyages in the 1860s and 1870s was William Carpenter of the University of London; and in March 1871 he gave a Discourse at the Royal Institution on recent scientific researches in the Mediterranean, in HMS *Porcupine*. But the acme of scientific voyages was that of HMS *Challenger*. She sailed around the world between 1872 and 1876, under the scientific direction of Charles Wyville Thomson, and was equipped with a chemical laboratory and many other up-to-date scientific facilities. This was one of the voyages in which the men of science were in charge, in that the point of the expedition was scientific; often there had been frustration among scientists on voyages because the captain's instructions, or his interpretation of them, did not let him stay at interesting places as long as they would have liked, or put enough boats and crews at their disposal. The results of this voyage were published in fifty large volumes between 1880 and 1895, and represented international collaboration in ocean-ography and marine biology; the most eminent authorities in the world were chosen to write up the various results and describe the collections.

This formal writing-up took nearly twenty years: it is one thing to collect data, and another to evaluate it and make it into public knowledge. But even before the *Challenger* was home, her voyage was being described and discussed in Discourses at the Royal Institution. In March 1874 Carpenter talked about the temperature measurements made in the Atlantic at depths down to about 2,500 fathoms and their implications. In January 1875 Huxley spoke of the recent work on the ship and its relevance to geology, he was particularly interested in the way in which debris accumulates at the bottom of the sea because this would help in the interpretation of fossils. The more directly comprehensible parts of the *Challenger's* programme were thus made available to the public at a provisional stage.

Debris at the bottom of the sea was coming to include not merely the remains of organisms perhaps on their way to becoming fossils, but also rubbish from man's activities. The scientific age, with its new chemical and metallurgical industries, produced more pollution and spread it about more than any previous epoch. Rivers in great cities became foul partly as a result of the beneficent introduction of the water-closet; for diluted sewage was thus discharged straight into rivers, whereas before it had been collected as 'night-soil' in concentrated form from privies and used on the land. The Thames in London became so opaque and smelly as to be a national scandal, especially when in 1849 and then in 1854 John Snow showed the statistical connection between drinking dirty water and catching cholera. Water from Lambeth was much safer than water from Southwark further downstream.

Better sewage systems, and in London the building of the Victoria Embankment to narrow the river and cover the pipes, could at least get the nastiness out of cities; and by the end of the century filtered and chlorinated water was being supplied by the water companies. Industrial pollution was a more difficult problem. In February 1875 Edward Frankland, an eminent chemist, gave a Discourse on river pollution. He mentioned the Clyde as polluted by sewage; Dighty Burn, near Dundee, by fibre manufacture; Red River in Cornwall, by tin mines; and the Sankey Brook in Lancashire by chemical works. He remarked that the River Aire, even before it reached Leeds, had received the house drainage of a quarter of a million people, and the refuse from 1,341 cloth and woollen factories, 1 silk mill, 1 flax mill, 10 cotton factories, 7 paper mills, 26 tanneries, 13 chemical works, 8 grease works, 4 glue works and 35 dye works. It need not surprise anybody that Victorian cities were unhealthy places. Royal Commissions had been appointed, and had recently reported, on the state of rivers; they had investigated the 'comfortable doctrine' that in twelve miles or so a river however foul would purify itself, and found it untrue. Dissolved organic matter, even in well-aerated rivers, was hardly reduced. Even in rivers less polluted, when some living organisms could survive, the water failed to purify itself. So far chemical methods had failed to produce purification, and it had also turned out that matter precipitated from dirty water would not make fertilizer of great value. Sewage works were not going to pay their way, or make a profit.

Frankland's answer was to use surface attraction rather than chemical affinity as the agent of purification; in particular, to purify by means of the soil, in irrigation. Near Aldershot, where this was being tried, one could see the drainage of London being transformed thus into strawberries and cream. Air and water brought alternately into contact with earth did the trick. The commissions had recommended 'mild' legislation about industrial pollution; and Frankland believed that if firmly enforced it

> would have the effect of restoring many of the rivers to nearly their original purity. To make them pure for drinking purposes is, perhaps, impossible; but it may reasonably be hoped that they may become sufficiently so to delight the eye and to repress the pestiferous and sickening exhalations which at present affect the multitudes of our population compelled to pass their lives on the bank of such rivers.

These were not new problems, though they had been made

worse by the rapid growth of population, industry and cities. In the early years of the century those, including Davy, who went salmon-fishing were concerned about the state of rivers. By the 1870s there was also genuine concern for the poor people who lived by or drank from them. The spirit of laissez-faire which some see as the root of Victorian values was giving way to legislative control in fields such as pollution. The hope expressed by Harrowby twenty years earlier, that the spirit of science might be imported into politics, is to some extent realized in Frankland's Discourse. What this spirit brought was a confidence that problems can be solved, and do not just have to be lived with. It was not yet clear that in solving practical problems as in solving those in pure science, one thing leads to another. There are no final solutions. This makes science a never-ending quest, and technology a matter of solving the problems posed by earlier solutions.

Harrowby had expressed faith in scientific education and in the scientific spirit generally, and these things were prominent in the Discourses at the Royal Institution in the 1870s. Indeed it was at this time that scientism really began to become important, not only in Britain but throughout western Europe, the USA, Russia and Japan. Belief in evolution, as a cosmic guarantee of progress, was a part of this scientism; but the physical sciences were just as important, because of the support they seemed to give to a reductionist philosophy. Deterministic laws of atomic arrangements in the triumphant years of classical physics seemed to lie behind the phenomena of life. In the next chapters we shall look at how scientific education, tending to dogmatism in its elementary stages at least, and the rise of physics as the leading science, affected the way people saw the world in the last decades of the nineteenth century.

9
Battle of Symbols and Jargon

The language of science overlaps with that of ordinary life, but words like 'field', 'elementary' and 'family' came to be used in the later nineteenth century in rather different senses in physics, chemistry and biology. Learning science is in part learning a language. As scientific courses proliferated, so this aspect became more prominent; the people working in particular sciences came to expect of each other that they would speak the same language, and gaps between physicists, chemists, biologists, geologists and so on increased. Writers in our day sometimes refer to 'the scientific community', but this is very much a theoretical entity: there may be some issues on which all active scientists feel the same, and are opposed to those in the humanities, but there must be very few. The real communities are discipline-orientated; and as science has grown in the exponential way we are now familiar with, so the scientific communities have become more specialized. Nobody could keep up with all the chemists nowadays, for example; though someone might from time to time attend enormous congresses of chemists, he or she would go to papers devoted to his or her special branch of the subject.

Scientific language is theory-laden: in using it, one is taught to see the world in a particular way. This happens to some extent with ordinary languages: in Malagasy one has to make it clear whether a 'we' is inclusive or exclusive, while in Navajo using the verb 'to go' requires one to give some clues about returning. Even apparently descriptive language involves selection of what are the significant features to describe, which will be affected by theory. A follower of Linnaeus' system would describe a plant a little differently from a devotee of the more natural methods of Ray or

Tournefort, for example. In the physical sciences the degree of abstraction and theory-loading is much greater. In a developed science, pure description if not impossible is very rare.

Victorian scientists were devoted to facts, and even more were the Gradgrinds for whom knowledge meant a multitude of facts, and who thought they knew what a fact is. In our world of competing newspapers and television, of leaks, press conferences and official communiqués, fact-finding commissions are necessary; and we know for example that there will never be full agreement on what are the facts behind the sinking of the *Belgrano* or the *Rainbow Warrior*. Facts are events which are both authentic and significant; in most cases we trust judges and juries, or less formal bodies, to determine them, but in complex cases it requires judgement and is by no means obvious. Anybody trying to bring about a major transformation in a science must direct attention to the facts as he sees them: to a herbalist in the tradition of Culpepper it was an important fact that a particular herb was collected under a waxing or a waning moon; for one of Liebig's pupils analysing it this was not a relevant fact at all. The derivation of the word 'fact' can indeed remind us that it means something constructed rather than just observed.

The physical science most intimately concerned with language was chemistry. Lavoisier had coined a new set of terms for it, replacing those which enshrined earlier theoretical views, from alchemy or from the phlogiston theory of burning. He hoped that readers of his book would find themselves willy-nilly thinking his way. He allowed himself to break his rule of a language based on observation by naming 'eminently respirable air' *oxygen*, which means the generator of acids; and this turned out to be a mistake. The hope for a language which would express the essence of things rather than being conventional noises perpetuating old-wives' tales is an old one, going back beyond Lavoisier to the seventeenth century. Then the decline of Latin meant that men of science in different countries could no longer understand each other. Latin anyway seemed ill suited for the new philosophy, although Linnaeus developed a new dialect of it, rich in adjectives, to provide concise descriptions and names of species of plants. To Lavoisier and his associates a new international language could be constructed for chemistry.

Mathematics had long had, in the 'Arabic' numerals coming originally from India, and in the symbols which had been invented from the seventeenth century on, an international language of symbols. The noises made when an Englishman a Frenchman and a German read aloud '2 + 2 = 4' are different, but to all of them it means the same. Chemistry had also had symbols, but the problem was that it had them in profusion. Alchemists had expressed chemical reactions and recipes in symbolic form, employing for example a sign used by astrologers for the gloomy planet Saturn to express its associated metal, lead. The difficulty was that there was no agreement about the signs, and down to the seventeenth century no strong feeling that there should be such agreement. Like other trades and 'mysteries', chemical or pharmaceutical knowledge was to be passed from father to son, or master to apprentice, or wise woman to her daughter, and not broadcast to the world. In particular in the sophisticated world of alchemy, the resonances of chemical and other truths were of the utmost importance. An alchemical text was not simply or mainly a means of conveying chemical knowledge, and certainly not of doing it succinctly. The process of making base metal into gold was akin to the maturing of a human being, and the texts were deliberately written to be read at various levels.[1] A dictionary of alchemical symbols is rather formidable, and indicates how many synonyms (full or approximate) there were, and how far this language was from being clear and distinct.[2]

The French reforms at the end of the eighteenth century were intended to end all this.[3] Chemical knowledge was to be communicated in a straightforward manner, leaving associations and word play to poets and writers of fiction like Goethe, whose *Elective Affinities* was one of the few novels to exploit chemical imagery with serious rather than comic intent.[4] The difficulty is that the world is so full of a number of things that there are many difficult names to be learnt; chemistry seems necessarily to involve a rebarbative vocabulary. Learning it is easier if it has some reason behind it; but such reasons are theoretically based, and will therefore date.

Lavoisier made the fundamental units of chemistry to be the elements, bodies which could not be further analysed; every reaction with an element yields a product which weighs more. By means of the balance – long used by assayers and druggists, but less by theoreticians – he made chemistry into something a little more quantitative than it had been. It became a science of more exact recipes, like a cookery book in which phrases like 'add a pinch of salt' or 'blend in enough milk to give the right consistency' had been replaced by terms indicating definite quantities. This was still some way from a mathematical science, and chemists continued to look for the Kepler or the Newton of Chemistry who would make it deductive and mathematical. Lavoisier's chemistry had a new language, but it had no system of symbols like the Arabic numerals.

Dalton in the opening years of the nineteenth century introduced the atomic theory, which gave a theoretical background to the ideas of definite proportions: if definite numbers of atoms of each element combine, then the total weights combining must also be definite.[5] If combination is atomic, then it should be possible to symbolize chemical compounds, and to set out chemical reactions in a kind of algebra. But Dalton wrote no chemical equations, though he did use symbols to indicate composition and perhaps even structure. His symbols were circles with different patterns inside them; presumably circles because it had always been taken for granted that atoms were spheres. They never caught on. They would be hard to remember; they would have been a nuisance to printers, who would have had to make a whole series of new pieces of type; but perhaps most important, they were ambiguous. Many of Dalton's circles had been used in the past in alchemy or pharmacy with one or more meanings, very different from that of one atom of a chemical element which Dalton meant them to represent.

There were all sorts of other problems with Dalton's atomic theory.[6] Earlier atomism had always involved atoms ultimately of the same stuff; and in the nineteenth century 'physical atoms', the real ultimate building-blocks of the world (probably of very few different sorts) were distinguished from Dalton's 'chemical atoms', the smallest units taking part in chemical changes. Dalton contributed to the muddle by writing of 'atoms' of compound gases like carbon dioxide. Compared to the discussion of the nature of matter in Paris or London, his theory was unsophisticated.[7] Its advantage was that it worked, in that it gave some hope of understanding why chemical compounds behave as they do; and it did open the way to symbolizing chemical reactions.

It was J. J. Berzelius in Sweden who in the second decade of the century put atomism on a much more secure footing. He used it to explain how different chemical compounds may contain the same elements in the same proportions – 'isomerism' – because their atoms are differently arranged, and how different substances may have the same crystalline form – 'isomorphism' – because they have the same number of atoms in the same arrangement. He also invented the symbolism we still use for expressing the atoms: H for hydrogen, O for oxygen, Fe for iron, and so on, based upon the initial letter or letters of their Latin name.

The difficulty was to agree on the proportions of atoms in particular compounds. We know that water is written H_2O, but this is not a fact which follows from an experiment or two. Dalton believed that the simplest compound of two elements must have one atom of each; which made water for him HO. Analyses showed that water contained eight parts of oxygen to one of hydrogen; on Dalton's assumption, then, the atom of oxygen weighs eight times as much as that of hydrogen, while on the H_2O formula favoured by Davy and by us it weighs sixteen times as much. There were various suggestive pieces of evidence, for example from the volume relations of gases – two volumes of hydrogen combine with one of oxygen, which suggests H_2O – but all of them posed difficulties of a theoretical kind. If Davy and Berzelius were right that chemical combination was electrical, the union of positive and negative atoms, then a combination of two hydrogen or oxygen atoms into H_2 or O_2 was impossible, for instance; and yet such union seemed to be required for the H_2O hypothesis. In the middle of the century there was no agreement on the formulae of even the simpler chemical compounds, and many chemists felt that an atomic theory was merely a distraction: that chemists ought to stick at the 'equivalent weights' which gave accurate recipes, and not bother with attempts at explanation.

One way out of this dilemma was to adopt conventions which led to definite formulae without worrying too much about how close they were to nature; and this was proposed by the French

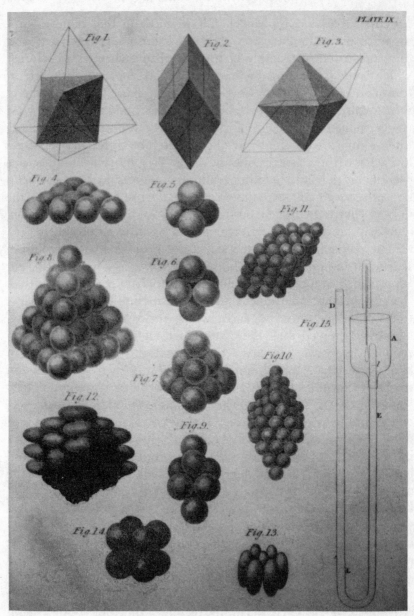

PLATE IX.

Figure 10   *Crystal forms built up from spheroidal atoms from A.*
Ure, Dictionary of Chemistry, *third edition, London, 1828, after*
W. H. Wollaston whose models are in the Science Museum,
London.

chemist Charles Gerhardt in the middle of the century. He organized chemical compounds into a series of 'types'; but for him formulae were simply condensed recipes, and a compound which could be made in two ways might well have two different formulae. His associate Auguste Laurent – and both of them were kept on the fringe of the French scientific establishment – was more hopeful: he believed that it was possible to determine formulae and structures, but not inductively as chemists had so far tried to do. Instead of hoping that experiments would eventually pile up evidence for some particular formula, he recognized that it was necessary to proceed hypothetico-deductively: to propose a structure for a given substance, then to work out consequences of this and test them. This turned out to be the way to go. Laurent died at the height of his powers in 1853, but his ideas slowly gained ground over the next ten years. His book *Chemical Method* was translated in 1855 by William Odling, a prominent figure at the Royal Institution who went on to become Professor of Chemistry at Oxford in the 1870s, as courses in science there became important.

Chemistry in the first half of the century was something which had to be learned by medical students, often without much pleasure; they had to have formal if elementary courses of lectures, and textbooks. Otherwise it was something to be put across in demonstration-lectures to those wanting to keep up with what was going on, but not to bother with too much detail. For them, something superficial would do; and the demonstration-lecture can also be dogmatic. After all, the demonstration was not merely something to watch, it was also meant to be demonstrative evidence, 'demonstrating' as the geometer demonstrates his theorem and puts QED when he has done it. The change with the founding of the Royal Institution (and then elsewhere) to lecturers who were active in research meant that audiences were sometimes exposed to questions rather than answers. They could see that there were problems presented in the work of Davy or Faraday; and if they were lucky they might in a later course see the problem solved, only as ever to raise more. Lecturers like these could afford the luxury of sharing their doubts and uncertainties with their hearers; and these make the most exciting and entertaining lectures.

Much science, however, is well-established, and some philosophers like Hegel have been most interested in this established part rather than in the frontier where conjecture, refutation and research generally is going on.[8] Many educationalists in the nineteenth century believed that for the young it was right to teach only what was certain, such as geometry and classical languages; once these had been mastered it would be all right to get on to more hypothetical subjects. When sciences like chemistry began to enter university and then school curricula, it was this sure foundation which their teachers expected to lay. Schoolteachers were not expected to engage in research, and by the 1870s this was anyway much more difficult than it had been in 1800: there was more to master first, and more equipment was needed. Science in this context was altogether more dogmatic than in Royal Institution Discourses; there was a syllabus to be got through, and right and wrong answers to questions at the end of it. Theories were valuable not as a guide to research, but as a prop to the young sprig: they were good insofar as they made a mass of information easier to handle and to remember. Teachers came to think of their job as akin to filling bottles, rather than to lighting fires.

The atomic theory thus came by the 1860s to have two functions: it might be a fundamental theory of matter, about which it was appropriate to argue in a very general way; or it might be a teaching aid, helpful to students who learned it as a dogma because it made sense of a great number of facts. At the Chemical Society of London there were great debates about the status of atoms in 1867 and 1869, with Professors at Oxford and at London Universities on opposite sides; and at the same time in France some of the most distinguished chemists were sceptical of atoms, among them P. E. M. Berthelot, who went on to become Minister of Public Instruction in 1886, having done distinguished work on synthesis and on calorimetry. He helped to keep atomism from being prominent in French syllabuses; but in Britain, atomic theory and the models with balls and wires which expressed it (then called 'glyptic formulae') became an important part of elementary science, whatever the doubts of the eminent.

Atomism made sense of the language of chemistry in that, if each symbol is understood as representing one atom, one can ask questions about how they are arranged; if the symbols merely represent numbers, equivalent weights, then everything is much more abstract. In 1860 the great international conference at Karlsruhe, the first chemical affair of its kind, voted that equivalents were more empirical than atomic weights; but ordinary mortals found it much easier to think in terms of particles, and to call them 'atoms', even though they might well be complex. There was an analogy here with Darwinian biology, where the species at a given time were real interbreeding units though they were all supposed to be descendants of one primitive form; similarly the chemical elements were supposed by many to be all 'descendants' or polymers of hydrogen or helium, and therefore not truly simple bodies, although in ordinary chemical processes they could not be transformed one into another.

Laurent's classic book of 1855 contained chemical equations like ours; writing for instance the reaction of potassium carbonate and sulphuric acid as:

$$H^2SO^4 + K^2CO^3 = K^2SO^4 + H^2CO^3$$

The carbonic acid immediately turns into H^2O and CO^2. Apart from having the numbers above rather than below (prompting jokes from mathematicians), this looks very modern. Not all Laurent's formulae are quite the same, however, as ours; those involving iron, for example, look different. With these we may compare the more primitive diagrams of reactions set out in an earlier and elementary work, J. B. Scoffern's *Chemistry no Mystery*, of which the second edition was published in 1848. He used no symbols, but tried to set out something like reaction mechanisms to make his subject more readily intelligible (p. 221):

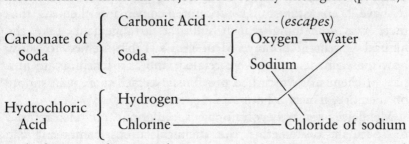

Such diagrams do not indicate quantities as our equations and Laurent's do, and they assume what Laurent denied, that chemical substances are composed by the union of opposites – that there is soda and carbonic acid in sodium carbonate (Na_2O + CO_2 in our terms, rather than a unit, Na_2CO_3). But the diagrams have the advantage of indicating what might be going on. A diagram which is in fact theory-laden may be a valuable aid for remembering elementary information.

The great success of a teaching programme in chemical theory was the Periodic Table of Dmitry Mendeleev, first published in 1869. He had been trying to put inorganic chemistry into an intelligible form for his students at the University of St Petersburg, when (in common with various contemporaries in various countries) he realized that if the elements are set out in order of increasing atomic weight, they display periodicity: similar elements recur at regular intervals. His success depended upon the general agreement on atomic weights and formulae which had followed the Karlsruhe Conference of 1860; after the conference delegates had read the paper of S. Cannizzaro, and came to agree with its recommendations. By taking volumes and weights into account, definite formulae could be fixed on, beginning with water as H_2O.

Given agreed formulae and atomic weights, the Periodic Table was waiting to be discovered; Mendeleev's contribution was to stake his reputation on it, and to see its power. When the elements are set out in the Periodic Table, relationships between those vertically above and below each other, those to the right and left of each other, and those on diagonals become apparent. Mendeleev used these relationships to predict the existence and properties of undiscovered elements; and with the discovery of Gallium, Scandium and Germanium his predictions were found to have been astonishingly accurate, convincing chemists that there was something real behind the arrangement. Like the natural system in botany, Mendeleev's Table seemed to show how the elements really were related; and as a teaching aid in a part of chemistry which had previously seemed to depend simply on memory, it was a boon.

Mendeleev resisted evolutionary speculations, and also refused to speculate on whether the chemical atoms composing the elements were the same as the physical atoms imagined by natural philosophers, the ultimate building-blocks. Ever since

Lavoisier, chemists had kept their science distinct from what was coming to be called 'physics'; Thomas Thomson even declared about 1840 that it would not matter if chemists and natural philosophers had different theories of heat – each could use what worked better for them. This would have led to science becoming a babel; but it shows the widespread fear among chemists that their science would be 'reduced' to something else by armchair theorists who did not know how to hold test-tubes or blow glass T-pieces. Chemists were even suspicious of physical methods of analysis; but the convenience of the spectroscope meant that by the 1870s it had come into regular use, as we shall see later.[9]

Chemical symbols and the atomic theory took those who used them behind, or perhaps away from, the phenomena; particular observations are only interesting in developed science against a background of abstraction and theory. For Lavoisier the crucial principle behind chemistry had been the conservation of matter. All chemical reactions were in a sense an expression of this principle; and if it seemed in some case that the weights of the reactants and the products were unequal, then the chemist must have missed something. In mechanics there were interactions in which momentum (mv) was conserved, and others in which *vis viva* (mv^2) was unchanged; but down to the middle of the nineteenth century there was no principle of the same generality as that which Lavoisier had made into the foundation of chemistry. With the proclamation of the principle of conservation of energy by Helmholtz in Berlin from 1847 and then in a public lecture delivered at Königsberg on 7 February 1854, the whole range of physical sciences began to take a new shape.[10] The great period of 'classical physics', the second half of the nineteenth century, depended on the confident extension of this principle to as many phenomena as possible. What had been separate sciences were thus unified. On the intellectual front, this unification went on at the same time that in the social history of science specialization was the order of the day.

Just as Newton was not the first to notice that what goes up must come down, so there were many groups of phenomena where conservation had been recognized long before Helmholtz. Indeed about a dozen claimants, from several countries, could be put forward as the discoverers of the principle.[11] At the beginning of the century, heat, light and electricity were separate sciences, and so was magnetism, which had connections chiefly with navigation. Heat had intimate links with chemistry, and optics with astronomy. While Galvani showed the connections of electricity with nervous impulses, Volta, Davy and Berzelius, as we saw, demonstrated its links with chemistry. The close relationship of light and radiant heat was discovered at the turn of the century in infra-red radiation, and that between light and chemical affinity in the ultra-violet rays which will set off reactions. Some philosophers saw 'force' as more fundamental than matter, and Schelling argued that polar forces underlay all phenomena, and that apparent rest was really equilibrium.

All this though was some way from any quantitative expression of conservation of 'energy' – a term first used in something like the modern scientific sense by Thomas Young at the beginning of the century. But whereas these ideas of the 'correlation of forces' led to insights like those of Oersted and Faraday in electromagnetism, they did not lead to the restructuring of the various sciences. The same was true of James Joule's experiment in which water was warmed quantitatively by stirring it in special apparatus: a definite quantity of mechanical force in the clockwork produced a definite quantity of heat – but the context of this was argument about the nature of heat, and Joule seemed to have proved that it was motion of particles rather than a substance. Helmholtz generalized what others had suggested, or had established in one context. By arguing that the dimensions (mass, length and time) in which the various forms of energy can be expressed must be the same, he opened up a whole new research programme. Electrical and magnetic units, for example, could be put into centimetres, grams and seconds; and doing this kind of work kept physicists busy for many years.

The science of physics really came to birth during the nineteenth century, though the process of convergence of separate physical sciences had perhaps begun when Galileo united celestial and terrestrial dynamics. In France at the end of the eighteenth century, there had been a science of 'experimental physics'; Thomas Young also spoke of 'physics' in a famous course of lectures at the Royal Institution published in 1807, but his physics included animal and vegetable life, and excluded optics and mechanics. That is, physics in the early years of the nineteenth century meant those parts of science which had not become mathematical; it occupied a place somewhere between natural philosophy and natural history, alongside chemistry. The word comes straight from the Greek meaning 'nature'; and it did not acquire its narrower modern sense until about Helmholtz's time. In France, one landmark was the election of the mathematician S. D. Poisson to the Académie des Sciences in 1812, in a 'physics' position.[12] In France, theoretical physics dominated by those with mathematical training came sooner than elsewhere.

Conservation of energy gave a good reason for putting all those sciences which dealt with energy in one form or another, and even those perhaps which dealt with its interconversions, under one umbrella. Chemistry and physiology have, however, generally been keen to keep out, even though it be in the rain. Electricity, magnetism, optics, heat and mechanics now all hung together; and whereas Dalton had simply concentrated upon weights in chemical reactions, the new science of physical chemistry came into being to study the energy and the rate of chemical changes. Conservation of energy was comparable to evolution as an organizing principle of the sciences; it opened new doors, and indicated new relationships.

The scientist who proposes new interpretations or principles is rather like the fictional detective (they may exist in reality too) who is faced with the same evidence as the reader and the police, but sees that it forms a different pattern when seen from the right angle. Conservation principles depend upon the idea that to explain change it is necessary to fix upon something constant. Atomic theory explained chemical change as the rearrangement of unchanging atoms, and therefore in a sense as superficial. Conservation of energy meant that the production of electricity from mechanical work, and ultimately from a chemical reaction (as in a power station burning coal or oil) is for example simply a transformation. The creation or destruction of atoms or of energy was held to be impossible, at any rate for mankind – the beginning and end of the universe was generally accepted not to be the concern of scientists, down to very late in the century. Science was essentially the study of the world as a going concern, not of its origin or end.

It would be very hard to prove a conservation principle. It is the sort of thing known by its fruit; and by the second half of the nineteenth century conservation of matter and of energy were such effective assumptions that they were generally taken for established truths. Evidence was reinterpreted that might seem to count against them: for example, the Earth's long geological history, which seemed impossible if it were just a cooling sphere; or the Sun's apparently endless emission of energy, which was hard to explain in terms of its being made of coal, or fuelled by meteors, or contracting and giving out gravitational energy as heat and light. In the event, these anomalies turned out to be due to radioactivity; but the point is that conservation principles are not readily falsifiable. They are the kind of assumptions which make whole sciences possible and rational, and they cannot therefore be tested within these sciences.

The principle of conservation of energy is sometimes called the First Law of Thermodynamics. This is slightly curious, because a 'law' in science had come to mean some relationship which had been derived from experiment.[13] It generalized and perhaps idealized a mass of experimental data, as Boyle's Law of the volumes and pressures of gases, and Ohm's Law relating potential difference and current through a resistance both do. Thermodynamics by contrast is a highly mathematical branch of physics, concerned with relations of heat and work, and conservation of energy really functions in it not as a law but as an axiom: as one of the bases for the deductive, mathematical system.

The meaning of the term 'law' in science has changed greatly over time; and during the Age of Science it was a term of great importance. Men like Gay-Lussac sought laws rather than risking theories; and Davy declared (*Works*, VII, 93ff.) that Dalton's Royal Medal from the Royal Society was for his laws of chemical combination, rather than for his atomic hypothesis.[14] In the Renaissance the word had implied that God had chosen certain laws for his world out of the possible range, like the founder of a Greek city-state deciding upon certain laws rather than others. God's laws of nature were like those of the Medes and the Persians, which changed not; but they were not necessary laws which could not have been otherwise. Indeed part of the drive towards modern science in the West was the search for the laws which God had in fact chosen to apply to the world; experiment as well as deductive reasoning were necessary to find them out.

By the nineteenth century the reign of law was an attractive idea to those inside the sciences and out; but the laws of nature had come to seem not merely unchanging but also inevitable. The 'possible worlds' which had interested earlier thinkers now seemed realized in other societies, as history and anthropology became developed; but in contrast to human societies, nature seemed fixed and uncompromising. A law of nature was the last word; and just as God could not break a law of logic (by making something both red and green all over, for example), so it was felt that He could not break a law of nature. The age of miracles was past; and indeed whereas in past times claims to miracles would help a religion, by the mid-nineteenth century they were a hindrance to it, connecting it with superstition.

What is interesting is that a different interpretation was also forthcoming, both from a lawyer and from thermodynamics. Henry Maine published his classic *Ancient Law* in 1861. He sought to show that legal conceptions are the product of historical development; but in particular he indicated how law functions as a framework within which we move. Thus the Romans were not allowed to sell their ancestral real property, and yet houses and land regularly changed hands through a system of legal fictions and collusive actions. Law was not absolute. In the same way, the laws of thermodynamics laid down a framework within which engines and refrigerators operate. The reign of law allows a fair range of liberty.

It was the Second Law of Thermodynamics which gave Victorian intellectuals a certain *frisson* because it seemed to imply that the world was running down. This proposition had been derived by Sadi Carnot in France in the 1820s in an investigation into the efficiency of steam-engines.[15] Steam-engines had been enormously improved since their invention about a hundred years earlier, and Carnot was interested in whether there was a limit to the improvement. He envisaged an idealized model of perfect gases and frictionless cylinders, in contrast to the engineers who had actually been making the improvements to real engines; but his method allowed him to demonstrate that heat will not flow spontaneously from a cool body to a warm one, and that the efficiency of a heat engine depends on the difference of temperature between the 'source' and the 'sink' of heat. Using steam under pressure (and therefore hotter), and cooling the condenser, would make a steam-engine work more efficiently. Engineers already knew this, just as we know that the refrigerator will not work if it is not plugged in; but Carnot's achievement, when recognized after a quarter of a century, allowed theoretical understanding and in the end further advance in practice. What is curious about Carnot's work was that it was done using the obsolescent theory that heat was a weightless fluid, driving a steam-engine like water driving a water-wheel. With this assumption his principle followed fairly simply, and later physicists – William Thomson and Rudolf Clausius – had problems in incorporating Carnot's law into the new theory, where heat was motion of particles and was actually being converted into mechanical work and not just flowing through the engine. They saw disorder steadily growing, like a baby, and called this entropy.

The Second Law of Thermodynamics thus constrains the designer of engines; he wants high efficiency but he also has to think about speed and price, and a whole range of possible designs might suit a particular case: the law does not require a unique solution. In 1859 at the British Association meeting in Aberdeen, James Clerk Maxwell suggested that the law was not absolute but statistical:[16] it described probabilities.[17] If a mass of hot and a mass of cold gas were brought together, then it was very highly improbable that they would not mix and give a uniformly warm mixture; but there was a small probability, and Maxwell imagined a demon who might assist the process, that the warm gas might separate out into a hot and a cold portion, in defiance of the Second Law. Darwin's theory had been criticized as the 'law of higgledy-piggledy' by Sir John Herschel, the eminent physicist; but in the same autumn, Maxwell had begun the process of turning physics too into a statistical science. This was not obvious to most people in the later nineteenth century, and the general effect of the Second Law was to encourage the gloomy belief that in time the universe would die the heat death of general tepidity: that the universe would see Victorians and their foreseeable descendants out, its death being many million years away, was little consolation to believers in progress, who saw it all stopping in the distant future.

Carnot's work was 'pure' science, whereas that of contemporary engineers such as George and Robert Stephenson was 'applied' science, to use terms characteristic of our day. But in the age of science this distinction was not generally made. 'Arts', or crafts, relied on practice; and these traditional activities were passed on from father to son, or at least master to apprentice, as Faraday learned bookbinding, and as Davy had begun to learn medicine. In contrast to *practice*, which was acquired by doing it, *science* was formally taught.[18] Science was public knowledge and not a craft mystery – religious 'mystery' too is something which makes sense in practice. 'Pure' sciences in the first half of the century were those without an empirical component, such as mathematics and logic; everything else, whether immediately useful or not, was 'mixed' science. The expectation was that when the laws in the mixed sciences had been found, something useful would follow from them; Benjamin Franklin's remark about a new discovery, 'What is the use of a baby?' was widely quoted.

Thus Davy was happy at the Royal Institution to give series of lectures on agricultural chemistry, and to work on tanning as well as on potassium and chlorine. His work had a practical impact in industry, and also affected attitudes to materialism. With his most famous work, the miner's safety-lamp, it seemed that science was saving lives as well as boosting the economy. At just the same time George Stephenson also invented a safety-lamp very like Davy's; but he did not know the principle on which Davy had designed his, and his lamp was the fruit of practice, of trial and error, guided by the general idea of restricting the supply of gas-laden air. Such a device could not be trusted, and it was the hope of scientists that in place of rules of thumb properly scientific methods could be widely taught. That way the industrial decline of Britain, deplored through most of the century by Presidents of the British Association and eminent men of science generally, would be halted. The Royal Institution had begun with an abortive attempt to train artisans in elementary science; mechanics' institutes provided lectures for what seems to have been chiefly a membership from the lower middle class. Like the socially superior literary and philosophical societies, they laid on a programme with much general culture rather than formal science courses.[19]

The same impulse which led to the Great Exhibition of 1851 also led to the founding of the School of Mines, following the French, and of the Royal College of Chemistry, which was intended to bring Liebig's methods to Britain. These institutions, on a site bought with the profits of the Great Exhibition, eventually formed part of Imperial College, London. Prince Albert was an enthusiast for chemical teaching in London, and A. W. Hofmann (famous for research on the amines) was brought from Germany to be the first professor at the Royal College.[20] In 1849 Hofmann, with Warren de la Rue, began translating Liebig and Kopp's *Annual Report on the Progress of Chemistry*; this was essentially the production of a number of Giessen professors, who under Liebig's leadership had made this small university one of the best-known abroad. The *Report* was intended as a review, giving a complete survey (according to its prospectus) of Chemistry and its Allied Sciences; it was to come out in the first half of the year following that reviewed; and it would give a faithful and 'whenever necessary, a complete digest of each investigation' in chemistry, and its applications in pharmacy, arts and manufactures. Physics, still seen as experimental rather than mathematical, geology and mineralogy were also to be included; and questions requiring confirmation, or more attention, were to be noted. The attempt would be made to 'expose in a lucid manner the relation existing between recent enquiries, and our actual store of knowledge', and not 'merely isolated and unconnected abstracts of the several individual investigations'.

This modern-sounding and research-oriented publication ran to only three volumes in its English form: Hofmann's aims and moderate success in London reflected those of the *Report*. The intention was that the College should benefit the economy, but this could only happen as science replaced practice, and as those with a real knowledge and love of chemistry began to apply it. It was always one of Liebig's convictions that the British expected practical results from science too soon. The greatest discovery at the College in the early days was that of mauve, the first aniline dye, by W. H. Perkin in 1856 during an attempt to synthesize quinine. It is an irony that despite Perkin's setting up a dye works and doing well out of it, the dye industry had become almost completely German by 1914. A crash programme was needed to make dyes even for Army uniforms. The reason for this seems to have been that Victorian British industrialists looked for a quick return, while the Germans were more prepared to take a long view. Hofmann returned to Berlin, to the most prestigious chair of chemistry in Germany and thus in the world, in 1865, somewhat disappointed at the slow development of chemistry in Britain.

Despite some disappointments, the general progress of science was by this time undoubted, and a source of pride even to those not connected with it. In all directions it seemed that scientific investigation was triumphing over ignorance, and scientific analysis replacing rule of thumb. It is to these triumphs, some of which could be made intelligible to laymen while others remained obscure, that we shall now turn, looking first at the discoveries rather than at their reception outside the scientific community.

10

The Triumphal Chariot

Those who held forth in the Royal Institution in the second half of the nineteenth century, or who addressed large audiences at the British Association or its equivalents in other countries, were confident; and they had reason for their confidence. The enterprise which had begun to take off in Napoleonic Paris had undoubtedly made enormous progress; far more was known about the world than had been a generation or two earlier. There had been upsets, rows, false starts; but if the progress had been sometimes dialectical, zig-zagging through arguments, rather than straightforward, it had undoubtedly happened.[1] Moveover, the results of science in the form of gas-lighting, electric telegraphs, synthetic dyes and railways were there for all to see. Science was no longer a programme; it was a great body of knowledge, parts of which had already shown themselves to be power. It is not surprising that there were many attracted to the idea that scientific knowledge was the only true kind, or at least that reasoning like that used in the sciences could solve all human problems. Belief in progress was not confined to supporters of scientism, but the two went very easily together; and we should now review this progress.

In the eighteenth century, the centre of science might have been the cabinet of curiosities, assembled by the dilettante. This might include objects from the three kingdoms of nature, apparatus like a globe and a microscope, ethnographical material, and books and illustrations to complement these various things. For research purposes, the laboratory was just beginning to be separated from the kitchen; but for the nineteenth century the centres were the laboratory and the museum. Davy at the end of his life remarked that the service of the laboratory was a service

Figure 11 A chemical laboratory from S. Parkes, The Chemical Catechism, *1822, frontispiece (reproduced by kind permission of The Royal Society).*

of danger; that few chemists could expect to retain a quick eye and a steady hand for very long, and would have to rely on the hands and eyes of assistants. There were certainly explosions in laboratories in Davy's time, and Faraday's chance came when his master was disabled after working with nitrogen trihalides; here they did wear goggles, but in general precautions were few. Those working with lead, like Fraunhofer making optical glass, or with mercury, like Faraday doing electrical work, probably suffered from heavy-metal poisoning. Fume cupboards for those working with gases and solvents only slowly came into laboratories.

By the later nineteenth century, these dangers were much less; and perhaps more striking is Davy's other remark, about how in his lifetime chemistry had ceased to be a science of furnaces and large quantities, and had become an activity to be carried on with spirit-lamps and test-tubes in the drawing room. It was not quite the world of the children's chemistry set; but Davy was indeed able to take a portable chemical laboratory with him, even on his honeymoon, in the form of a box of apparatus. In the 1850s, the invention of the Bunsen burner meant that very high temperatures could be achieved on the laboratory bench; and if the furnace was no longer necessary, then the laboratory could move upstairs. The drawing room was not very convenient, because the chemist needed services such as water and gas; and the laboratory with its wooden benches to work at, with their sinks and gas taps, high stools to sit on, and racks of bottles at hand, came into existence. Chemistry was *par excellence* the laboratory science of the earlier nineteenth century, but experimental physics and physiology needed similar facilities.

From the seventeenth century, there had been specialized instrument makers; and in the eighteenth century James Watt was one, who turned his mind to bigger things. But it is striking in Faraday's *Chemical Manipulation* (1827) how much the scientist had to do for himself. Faraday's advice was never to throw anything away; and most of the apparatus he described could be home-made from glass-tubing and sheets of india-rubber. Filter papers were cut out of sheets of 'bibulous paper'; and complex distillations, such as that in which Faraday isolated benzene, could be done in zig-zag glass tubes. The chemist had to be a craftsman and a cook as well as a thinker; only the chemical balance was beyond the capacity of the resourceful scientist to make. But Faraday's puritanism, and his skills acquired in his father's smithy and his own bookbinding apprenticeship, were not typical; and entrepreneurs like Frederick Accum were already selling ready-made apparatus to those who preferred to buy it. Throughout the century, the quantity of apparatus commercially available greatly increased.[2] While the researcher might well need to make some of his own, standard equipment would be bought – partly for convenience, and partly because like the balance it would be more accurate for quantitative work then anything home-made. There may be some who find it relaxing to make some test-tubes or cut out some filter-papers; and apparatus to investigate previously unsuspected effects must be made from scratch, or assembled opportunistically from things designed for old purposes; but ready availability of apparatus is an enormous bonus to the scientist, and the growth of science led to a flourishing trade.

Parkes' *Chemical Catechism* of 1806 had a frontispiece of apparatus, engraved on glass with fluoric acid; in the fourth edition (1810) this was replaced by a plate of the laboratory of the Surrey Institution, a less-successful version of the Royal Institution; in later editions Parkes substituted a laboratory which a gentleman might copy for himself. In the early years of the century, laboratories were for one or two people, with assistants and perhaps a friend or two to watch, to perform research; or else to carry out routine analyses for a fee; or a mixture of these things. Those at lecturing institutions would also be used for preparing demonstration-experiments. It was not until the 1830s that the idea of a teaching laboratory became at all general; at Durham University then, for example, laboratory instruction was still an optional extra, to be paid for if taken. By the second half of the century, the teaching laboratory had become a feature of any university, especially in Germany, which tried to attract an eminent person like Du Bois-Reymond or Carl Ludwig by promising him an 'institute' with research and teaching facilities.[3] The great man might expect to play a major part in the design of the building,[4] as Maxwell did when the Cavendish Laboratory at Cambridge was built following his appointment in 1871, as the first professor of experimental physics.

Faraday had begun as a chemist,[5] and indeed was regarded as one by most contemporaries throughout his life. But even in the 1820s some of his work had been on electricity and magnetism, which would now be thought of as physics; and in the 1830s he moved decidedly in this direction.[6] The greatest authority in electrical science was A. M. Ampère, who had devised equations to account for the discoveries of Oersted and others, and was thus bringing electromagnetism within the realm of applied mathematics. But the problem remained that the physical basis of Ampère's equations was implausible; they provided a model which could be adapted to fit new discoveries, but which did not lead to new predictions or give a convincing explanation to those whose first language was not mathematics – like Faraday. Faraday seemed to contemporaries an experimental genius, thinking with his fingers; he was thus a splendid example of Baconian philosophy in action. They duly applauded when he quantified electrochemistry in his famous laws of electrolysis, and when he proved experimentally that electricity from a condenser, a battery or an eel was the same.

As Faraday went deeper into electromagnetism, he began to leave his contemporaries behind; his work was on the boundary between inductive experimental science and deductive physics, and failed to fit easily into either category. As he wrestled with the nature of matter and the question of action at a distance, he seemed to be going into metaphysical speculation; he was not a clubbable man, and his scientific work in its higher reaches was a lonely business, interrupted by a major breakdown in health. In the laboratory he did undoubtedly think with his hands thoughts which he could not clearly express in words, or in the more accurate language of mathematics (which he distrusted); but he did try in his lectures to put across some of his underlying convictions. This was especially true when he gave an impromptu lecture on the nature of matter in 1844 – the story is that Charles Wheatstone panicked and fled, leaving it to Faraday as chairman to satisfy the audience – and another on ray-vibrations in 1846. Both these are very different from the 'experimental researches' which made up Faraday's main research-publications, where every paragraph is numbered consecutively up to 3,299, and where theorizing is kept in its place in the Baconian manner.

Where most saw atoms with void space between them in the manner of Newton and Dalton, Faraday by the 1840s believed that matter was made up of mere point centres of force.[7] It was forces and not mass which he perceived as crucial; and thus the field between the points was where the interest lay. He saw it as filled with lines of force, whose arrangements permitted or prohibited the passage of an electric current; this was itself force rather than any material 'juice'. He believed that the ray-vibrations invoked to explain light and radiant heat might happen in the lines of force (which he seemed to have almost demonstrated experimentally) rather than in the hypothetical ether.[8] He declared: 'I do not perceive in any part of space, whether (to use the common phrase) vacant or filled with matter, anything but forces and the lines in which they are exerted.'[9] This led him towards ideas of conservation of energy, and of a unified field;[10] he believed that light and magnetism must act upon one another, and to the astonishment of contemporaries he indeed demonstrated that a magnetic field will rotate the plane of polarization of polarized light. Wheatstone's work in telegraphy seemed to show that electricity travelled about as fast as light, and Faraday believed that it must go just as fast; he also believed that gravity must be analogous to other attractive forces, and take time for its propagation, though there was no evidence for this.

Some of Faraday's experimental results were independently achieved, sometimes before him, by Joseph Henry in the USA,[11] and by William Sturgeon in Manchester, for example; and his theory of point atoms owes something to Newton, and more to Roger Boscovich (an eighteenth-century Jesuit) and to Joseph Priestley. The fertile combination of theory and practice was his own, however, and he stood at the head of a great group of people working on the new phenomena of electricity. William Whewell in Cambridge was consulted by him over nomenclature, and took a great interest in his work, especially in his unease about the atomic theory with its inert and massy billiard-balls. Similarly, important physicists on the Continent corresponded with him, and were amazed and delighted by his discoveries; but the lines of force seemed somehow unscientific, a scaffolding that could be ignored when the building was completed. He himself felt his lack of mathematics; but in science it may be an advantage not to be too well-drilled – one who knows the rules may find it hard to go beyond them, and where great originality is needed the prizes sometimes go to outsiders.

Two Scots with Cambridge connections took up Faraday's work at last, trying to put it into mathematical form rather than to fit the discoveries into an existing theory; and through their work came the great flowering of classical physics.[12] William Thomson, later Lord Kelvin, was elected to the chair of Natural Philosophy at Glasgow in 1846 at the age of twenty-two, and organized there a physics laboratory for students. He had already begun publishing on electromagnetism, and in a paper of 1847 he suggested an analogy between an incompressible solid and a magnetic field. In 1851 he used what were in effect vectors in deriving magnetic equations without using Poisson's theory of magnetic fluid. He went on to work in thermodynamics, and like Wheatstone to put his electrical knowledge to practical use in devices connected with telegraphy. He became the doyen of British physicists, and incidentally one of the first whose voice we can still hear, preserved in a primitive recorder at Glasgow University.

James Clerk Maxwell, who built upon Thomson's work in electromagnetism, was a younger contemporary, born 1831 and dying in 1879.[13] While at Cambridge he worked on the problem of the stability of Saturn's rings. In 1856 he was appointed to a chair at Aberdeen; in a series of reforms and cuts, he lost this job; but in 1860 he went to King's College, London. From 1865 to 1871 he lived on the estates he had inherited in Scotland, and then spent his last years at Cambridge. He was one of the greatest physicists there has ever been; we have already met his work on gases, with its statistical basis, but his most important research was that in which he took up Faraday's discoveries, giving the lines of force a mathematical expression. Beginning in 1855, and then especially in the early 1860s, he developed his theory of electromagnetism, which began with a model of tubes of force with complex mechanical analogies, and led to the idea that light was indeed an electromagnetic disturbance. In 1864 he published a book, *The Dynamical Theory of the Electromagnetic Field*, in which the equations were largely separated from these mechanical models; in 1873 he published his *Treatise*, which put the whole of electricity and magnetism on a new footing. His work was swiftly taken up in Germany, where it had close relations with that of Weber and Kohlrausch; and it was there that H. R. Hertz, for whom Maxwell's theory was no more than Maxwell's equations, demonstrated the existence of electrical or radio waves in accordance with the equations.

By then, not merely was much more known about light, heat, electricity and magnetism, but the knowledge was unified with a powerful theoretical structure, so that while chemistry had seemed the fundamental science to many at the beginning of the century, by the end of it there was little doubt that this position belonged to physics.[14] In Maxwell, the streams of particle and wave physics had begun to converge, with results that were to be disquieting; but optics and the structure of matter were at the same time fruitfully interacting. It had long been known that some metals gave colours to flames; but this was an unreliable test, because colours are hard to describe exactly, and because they are usually masked by a brilliant orange-yellow. Joseph Fraunhofer, an optical-glass maker of Munich, had mapped over five hundred dark lines in the solar spectrum by the early 1820s, and had shown that they did not arise in the prisms used to produce the spectra. One of the greatest of Cambridge physicists, George Stokes,[15] had obscurely suggested that elements might emit when hot and absorb when cold light of the same characteristic frequencies; but it was not until 1860, with the collaboration in Heidelberg of Bunsen and Kirchhoff, that spectroscopy became a central part of chemistry.[16] They realized that each element had its own spectrum, a pattern of bright lines (from which the ubiquitous sodium lines could be distinguished); and they used the spectroscope to detect two new metals, caesium and rubidium, which they were then able to isolate. Kirchhoff reasoned that cooler elements in the outer layers of the Sun would absorb energy according to the same pattern; and accordingly he identified in the Sun a number of elements found upon the Earth. As apparatus increased in power, elements were also identified in stars and nebulae.

The French philosopher Auguste Comte had said that among impossible things was stellar chemistry; and yet twenty years later he was proved wrong. Nineteenth-century philosophers were anxious to produce systems of classification of the sciences, and at the end of the century Robert Flint of Edinburgh wrote a *History of Classifications of the Sciences* (Edinburgh, 1904) which summarized them all. Nobody seems to have put astronomy near chemistry; and the trouble with all series is that progress often seems to happen on the frontiers of sciences that on the philosopher's map seem as far apart as Finland and Portugal. The coming of chemistry revolutionized astronomy as the coming of terrestrial mechanics had done in Galileo's day; it accelerated the move from positional astronomy towards physical astronomy, where the stars are not simply fixed points but entities which have evolutionary histories and form part of great systems.[17]

Stellar distances were first determined in 1838, when F. W. Bessel measured the parallax of 61 Cygni; this is the apparent motion in the star produced by the Earth's movement in its orbit, as predicted in the Copernican system. William Herschel at the beginning of the century had already detected the proper motion of the Sun within our galaxy. Bigger and bigger telescopes were used to investigate whether nebulae were star-clusters or stars in the making; this was something the spectroscope ultimately answered, because some nebulae gave bright-line spectra characteristic of hot gas rather than of the Sun or a star. Now the colour of stars, and especially whether they were red or white, became something of interest; and like natural history, the observation and recording of red stars was something which amateurs could do, and astronomical societies flourished throughout the century, bringing mathematicians and owners of telescopes together. At big observatories, there were elaborate programmes of observation, assisted in the second half of the century by photography. Especially in the USA, women began to play a role in observation in the big observatories by the end of the century.[18]

One of the great names in the new physical astronomy was Norman Lockyer, who rapidly mastered the techniques of spectrum analysis.[19] Having studied the sunspots and the solar prominences, in a race with the Frenchman P. J. C. Janssen, Lockyer then moved on to claim the existence of a new element in the sun, helium. He had seen its lines in the spectrum, and realized that they did not correspond to any known substance. This was in a sense the first of the family of rare or inert gases to be discovered; but in chemistry the discovery of an element requires also its isolation, and this did not happen until 1895, when William Ramsay was in the process of finding the whole group of these gases. Lockyer classified the stars, recognizing two kinds of history for them. He became eminent in the world of science, editing *Nature* and playing an important part in establishing the Science Museum in London.

This grew out of a Loan Exhibition (accompanied by lectures) of scientific apparatus for the 1862 Exhibition at South Kensington, which turned out to be a pale shadow of the Great Exhibition of 1851 – the death of Prince Albert adding to the gloom. By then apparatus was available for chemistry, physics, astronomy and biology; these ideas made in brass were not only useful but also often handsome. New apparatus made new

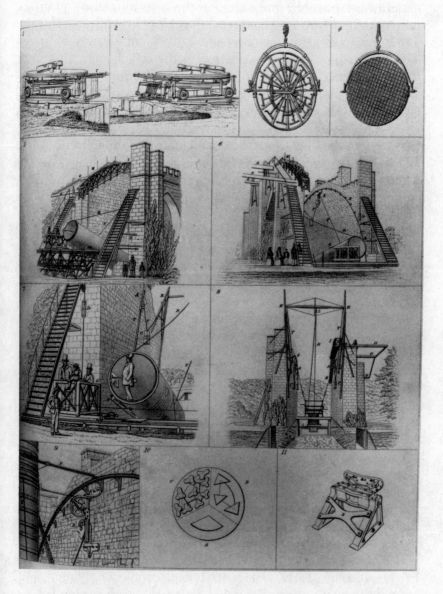

Figure 12   Lord Rosse's telescope from Philosophical Transactions, *1861 (reproduced by kind permission of The Royal Society).*

experiments possible; for example, the new diffraction gratings prepared by H. A. Rowland at Johns Hopkins University from 1882 led to a great improvement and quantification of spectroscopy, while the microscope led to the cell theory. Museums like the Science Museum now built up collections of enormous value to the historian of science, for the provenance of the pieces is known, and we can only judge past science when we know what could be done with the tools available. Apparatus known to have been used by a particular scientist, and collections of apparatus by the same maker, are particularly instructive.[20] Chemical apparatus tends to be broken, or reused in new arrangements; but the historian may hope that even here laboratories may have had capacious cupboards into which unused pieces have been stuffed and thus preserved.

The museum with old apparatus and machinery is one kind, but in the nineteenth century it was the museum as a centre of research in current science which was more important. In the late eighteenth century, John and William Hunter had built up collections of anatomical specimens for teaching purposes; the Hunterian Collection in London became one of the sights not to be missed by the intellectual tourist in the early Victorian period, when Richard Owen was in charge of it. As well as medical preparations, it contained a great deal of comparative anatomy, including fossils. This collection was in the care of the Royal College of Surgeons; and other medical societies and medical schools built up museums where the student could study at leisure, augmenting what he had learned from lectures and dissections and at the bedside.

Owen, the 'British Cuvier', had made the collection a centre of scientific research with his work, especially on fossils but also on the relationships between the various groups of animals. The model on which he and others based their plans for museums was that in Paris where Cuvier, Jussieu and Lamarck had worked in Napoleonic Paris. Here vast collections of plants and animals – stuffed, or as bones and fossils – were assembled so that comparative study could be carried on, and unknown creatures described and named. The arrangement of the exhibits was crucial: when W. C. Williamson went to Manchester to take charge of the museum there in 1836, he found that the stuffed animals were merely arranged in ornamental groups; he soon re-organized them systematically.[21] Everywhere the Parisian example was being followed, with differences in scale; the public's desire to be instructed in such museums, and also at exhibitions of appara-tus, is a striking feature of the time. The geologists soon joined in, and the Museum of Practical Geology in Jermyn Street in London, connected with the Geological Survey and arranging programmes of lectures, presented that science in a systematic manner too.

The cabinet of curiosities might be rudely described as a well-organized attic; and national museums are the national attic, where things too good to throw away are kept safe. The British Museum at the beginning of the century was an enormous cabinet where classical sculptures, ethnographic material and natural history specimens were all to be found. There was a mixture of the formal and the informal about it; the Keeper of Zoology in the 1810s was wont from time to time to leap over the stuffed animals. Following delicate negotiations with Robert Brown, the great collections made by Sir Joseph Banks were transferred to the Museum;[22] but the incompatible demands of the various departments, and the great library which also formed a part of the museum, led to a good deal of friction. Eventually Owen was appointed Superintendent of the natural history collections, and in 1881 presided over their move to the new building in South Kensington.[23]

Museums are part of an educational system, and part of the entertainment industry; and the problem was particularly acute in scientific museums, where botanical and zoological research was going on.[24] Just as serious students with a syllabus to get through needed lectures different from those whose hobby was science, so those studying biology or medicine needed a different sort of museum from the general public, excited by dinosaur bones and stuffed grizzly bears. Collections of dried plants are of interest only to the dedicated or desiccated botanist; and much the same applies to systematic collections of beetles or of fleas. Museums themselves came to need attics or basements where working collections would be available for research purposes. Nevertheless, museums were among the first scientific institutions to receive direct government support for science, and particularly for scientific research.

Most of this work was descriptive. Plants, animals and rocks collected at home or abroad – perhaps from professional collectors, like A. R. Wallace – came to museums in capital cities, university towns or other centres, where they were named and classified. We remember the great names like Lyell or Darwin, and forget the supporting cast without which they could not have been stars: those in museums classifying their findings. Darwin's finches,[25] for example, could be seen to be extremely interesting only when their affinities were pointed out to him by a taxonomist. Speculation about the origin of species began only when there was some agreement, and much argument, about what species were, and what their names were. This naming seems close to the boring and routine 'normal science' that Thomas Kuhn described in his *Structure of Scientific Revolutions* (1970); there is a good deal of routine in science, as in other things, but it is not always the most eminent who have done the interesting bits.

Lyell made people aware by the 1830s of the great age of the Earth, and he also divided up the Tertiary period, using a statistical method based upon the proportions of existing species found among the fossils. The great business of the geologists of the middle of the century was to settle the strata, making geological maps and connecting their relative dating schemes across the world. In 1859 Darwin could still write in the *Origin* of the imperfection of the geological record; only relatively small parts of the world's surface had been mapped by geologists. A good deal had not yet been mapped by anybody; not merely the interior of Africa, but parts of the western USA.[26] In the middle years of the century, the American government sent expeditions to explore the western territories and plot possible routes for the Pacific Railroad; a division of the Army, called the Topographical Engineers, played an important part in these surveys, which were then sumptuously published by the Congress. Work like this also produced much information about the distribution of plants and animals, and turned up interesting fossils: the pursuit of manifest destiny could also benefit pure science. Such surveys, whether by sea or land, again required not merely naturalists to accompany them, but museum staff to write up the collections and make the knowledge public.

Geologists had chiefly looked horizontally where fossils were concerned, using them to date strata rather than following resemblances vertically through time. Darwin's work gave new directions: first to fill in the gaps in the record so that the history of life would be more complete; and second, to investigate the origin of mankind, previously a tabooed subject. The first of these objectives was particularly pursued in the USA, where humans were relatively recent but where the fossil history of the horse seemed to be well displayed, and where there were curious links between reptiles and birds. The chief investigators here were E. D. Cope and O. C. Marsh, rivals in the manner of robber barons, chartering trains and suborning each other's assistants in an effort to get the spoils back to their base first. The American Gilded Age after the Civil War was a period of ruthless competition even in palaeontology, where the survival of the fittest in the struggle for existence was indeed worked out, in the present as in the past.

Darwin's theory was presented in the *Origin* in a curious way.[27] The book was an abstract of a work which had not appeared, and which never did in the form envisaged; this meant that it had no references or bibliography, and that in it Darwin wore his immense learning lightly. Its structure is like that of Paley's works: the argument is cumulative, but it is not a chain in which each link supports the next, as in geometry; rather, the various arguments are independent, and support each other like the fibres in a rope. The reader may not be persuaded by some of the cases, but the whole mass of them carries conviction. The book did not mention man other than obliquely, and it was not until 1871 that Darwin published *The Descent of Man*; but it was clear from 1859 that the ancestry of man rather than that of horses was the crucial question. Indeed, it had been the major objection to Lamarck's theory and to *Vestiges* that man was treated as an animal; and that if such a view gained currency, people would start to behave like animals.

From 1859 caves became exciting to geologists, as they had been forty years earlier when Buckland looked for evidence of the Deluge. Very painstaking stratigraphical work put it beyond doubt that men had been contemporary with animals now extinct, and that human history must therefore go back long before the 4004 BC computed from Genesis. The evidence came from Britain and from France, and was brought before the public when in 1863 Lyell published his book *The Antiquity of Man*.[28] These careful excavations were followed up by archaeologists, who were ceasing to be treasure-hunters, just as geologists had ceased to be fossil-hunters; both were now concerned with dating and provenance, and the reconstruction of past epochs. But just as man's dates were being pushed back, there came an unhappy interaction between geology and physics, again two sciences which had not seemed to have a common frontier. Post-Darwinian geologists had assigned vast tracts of time for the operations of forces still believed by us to have shaped the Earth; but they were challenged on the grounds that thermodynamics could demonstrate that the Sun and the Earth could not be much more than fifty million years old.[29] A confrontation between William Thomson and T. H. Huxley was good to watch, and showed that scientists did not always agree, even about what was evidence or the right way to argue: while evolution in the last thirty years of the century was generally accepted, there was no agreement about times and processes.[30]

Natural history had been based in the museum, and in the field; but the newer science of biology, named by a Frenchman and a German and prominently pursued by these two nations,[31] became increasingly a laboratory science with experiment rather than description as the key. The microscope, improving radically in the early nineteenth century so that its images no longer had coloured fringes all around them, was perhaps as important as the telescope had been two centuries earlier.[32] Cells had been noticed by Hooke in the 1660s with a primitive microscope; but the cell theory of Schleiden and Schwann gave a quite new understanding of how plants and animals work. Typical of microscopic work might be W. C. Williamson's work on the formation of bones and teeth in the 1850s, which got him into the Royal Society. He was a poor boy from Scarborough, who went to Manchester. He worked his way through medical school in London, returned to establish a good practice, and pursued scientific research in his leisure time, also giving public lectures – over 300 between 1874 and 1890 – and teaching at Owen's College, which became the University of Manchester. Exploiting a new device and a new theory, Williamson made a career in science for himself; and with later work on coal and the plants from which it had come, he made geology also a laboratory subject, involving apparatus for making thin sections, and a microscope.

The microscope was vital for physiology and for embryology, and later also for advances in medicine – notably the germ theory of disease. But during the second half of the century there came a new approach to physiology, involving chemistry.[33] Chemistry had always been a part of a medical training, with its role in pharmacy; but the organic chemistry which Liebig had pioneered promised deeper-level explanations in biology. Chemists had abandoned the belief that in organisms the balance of affinities was different from that in the inorganic sphere; they denied any vital force. Their incursion was therefore seen as materialistic; biochemists saw animals and plants as natural laboratories, and not as goal-seeking creatures interacting with their environment. The latter part of the century saw great progress in organic chemistry, as the atomic theory led to structural understanding and natural products such as indigo were synthesized. Particularly in Germany, there were close links between the dye industry and academic chemists; but in a new form, chemistry retained and expanded its medical connections. Physicists might be reducing chemistry to physics, but chemists seemed to be reducing biology to chemistry. The process which had begun with Lavoisier's study of respiration and continued with Prout's work on digestion led on to studies of proteins and fats. Physical chemists like Vernon Harcourt in Oxford were just beginning to study reaction-mechanisms in a few simple cases in the 1860s, and at the same period in France P. M. Berthelot was performing total syntheses, and in Norway C. M. Guldberg and P. Waage discovered the Law of Mass Action governing chemical equilibria. By the last decade of the century Emil Fischer was performing elaborate syntheses of carbohydrates and of peptides; and it was possible to suggest reactions and equilibria which characterized living things.

To the laboratory and the museum as a focus of science we should add the observatory. Observatories go back many centuries, perhaps even to Stonehenge; but modern observatories in the West began in the sixteenth century. By the nineteenth, they were very important and expensive institutions which had moved far from their original concerns with navigation and positional astronomy. Time-signals, geophysical measurements, weather reporting and the testing of chronometers were among the functions which they discharged. Experiments like those done in France on the velocity of light, by Fizeau and Foucault, and in Britain by Airy on the variation in gravity in a coal mine, were under the auspices of the national observatories. The making of great telescopes, and of the mountings which would direct them and keep them focused on a star, required great industrial resources and skill. It was at observatories that fundamental work on errors in observation were done by Gauss, leading to statistical theory applicable in all kinds of other fields.[34]

Some science was independent of all these institutions, or could have been: notably mathematics.[35] W. R. Hamilton was appointed Astronomer Royal for Ireland and a Professor while still an undergraduate in 1827.[36] He was no great observer, but a most original mathematician, making great advances in algebra (notably with the idea that xy need not equal yx), which had connections with optical theory, and then with vector analysis. His friend Augustus de Morgan was working in London on the foundations of logic, as was another Irishman, George Boole, who developed an algebra of classes. In the USA, Willard Gibbs, given a new position but no salary at Yale, worked on thermodynamics and the Phase Rule; and in Germany Georg Cantor developed the theory of sets. The explosion of mathematics was one of the most striking aspects of the nineteenth century; and frequently new mathematics has turned out, as Galileo had hoped, to be the language of nature, providing a key or a model for the understanding of phenomena.

What these mathematicians had in common with most of those who worked in laboratories was that the majority were attached to universities. The structure of universities in Germany, France, Britain and the USA was very different, but everywhere they were by the end of the century centres of scientific research, while at the beginning of it they had mostly been very marginal to it. German universities were under the control of the state bureaucracy, which appointed the professors, but they had responded quickly to the specialization of the period with the creation of new chairs.[37] The German professor was a powerful figure, hiring and firing his assistants and doing his best to place his favourite students, though by the end of the century this was becoming more difficult.[38] In France the system was much more centralized, and it might be that the stiffness of the Napoleonic system was one reason why France had lost to Germany the prominent place she had had at the beginning of the century when that system was set up.

In Scotland, the universities had always had strong scientific and philosophical traditions; but in England at the beginning of the century neither of the two universities was a centre of scientific research. Both were essentially seminaries – a majority of undergraduates went into the church in the first half of the century – and finishing schools, where teaching rather than research was the norm. The change by the end of the century was dramatic. The ancient universities had adopted German ideas, so that scientific research was now an important part of what went on, especially at Cambridge,[39] with physics and physiology.[40] Furthermore they had been joined by universities in commercial and industrial cities, beginning with London, and including Manchester, Birmingham, Leeds, Liverpool, Newcastle and Bristol. These were places where at the beginning of the century there was no more than a Mechanics' Institute or a Lit. and Phil.; and these new institutions were more oriented to science than the ancient ones, and committed on the German model to research.

In the USA, at the beginning of the century, the universities were also closer to seminaries, but by the 1870s they had begun to adopt the German model too, following the lead given by the new Johns Hopkins University in Baltimore, the first to emphasize research in alliance with teaching. By the end of the century, the national differences and traditions prominent earlier were less significant, especially in the physical sciences; and everywhere scientific education was becoming a serious and systematic business.[41] There was so much new knowledge in science, and degrees in science came to be seen as an alternative to degrees in classics or mathematics rather than as something to be done afterwards as an optional extra, or taken as a merely voluntary course of lectures. What is curious for the observer from the twentieth century is the debate about the possibility of secular education, something after all we now take for granted, although we do hear of debates over the teaching of 'creationism' in backwoods. At Oxford and at Cambridge, the new science degrees (like those in modern languages and history) led to a BA, and knowledge of Latin and Greek was essential for would-be students of chemistry or physics. This was because a degree was part of the training of a gentleman, and not a technical qualification; the scientist was as well-educated, though in a different line, as the classicist.

The utility of science was indeed everywhere urged, just as the utility of classics was urged for those who were to be legislators, justices of the peace or district officers in the colonies. Apologists for science also argued that doing science was valuable for building the character: accurate observing and truthful reporting were required of the scientist; and so was open-mindedness and humility in abandoning preconceptions. The feeling of sharing in a great enterprise for the relief of man's estate was also good; and the hope was that the method of science would prove applicable in politics, in the arts and in religion. Problems would be carefully defined, and the solution dispassionately searched for and no doubt soon found. Scientism is a not-unattractive doctrine, and was especially so to a rising professional middle-class who associated with it theories of eugenics and of mankind which gave them a pleasing sense of class and racial superiority;[42] but in the later nineteenth century there was no reason to anticipate these darker sides of progress.

One great difficulty in the event was that the picture of the man of science as an open-minded searcher after truth does not fit all of them; indeed it is hard to see how anybody could work in the kind of vacuum envisaged by some of those who invoke a Baconian inductive method. The great achievements of nineteenth-century science – atomism, energetics, evolution, the wave theory of light – were highly theoretical, depending upon conjecture and then search for evidence. And if research does not fit the pattern, teaching generally fitted it even less well. There were some, like Henry Armstrong at the end of the century, who urged a 'heuristic' method of teaching in which the pupil would be helped to make for himself the discoveries of Newton, Lavoisier and

Faraday. But the problem is that this requires a teacher of genius; and that a pupil has anyway only a brief time to get through work which has taken the lifetimes of many eminent predecessors: there must always be something artificial about heurism.

Most science was therefore taught dogmatically, as facts organized by proven theories, and as a series of hard words and mysterious formulae – the equivalent of learning an ancient language. While the experts knew that the status of the atomic theory and of facts and laws was uncertain, those learning elementary science absorbed a world-picture in which unchanging atoms of matter moved according to deterministic laws, in which useful knowledge was paramount, and in which problems could be solved by the almost-mechanical application of scientific method, which was really organized common sense. This was a view unsympathetic to religion and to the imaginative arts; but it was not characteristic of the leading scientists, who were often interested in all kinds of things beyond their science and continued to resist the appearance of two cultures. Thus Sir Archibald Geikie (of the Geological Survey) in an address at Birmingham in 1898 deplored the idea of splitting schools into two sides or sections; but the trend towards specialization proved hard to control, and to some the sciences did seem to be all that needed to be known. Indeed it seemed to some that scientific investigation might even bring to light all that was worthwhile in religion.

11
Wrestling with the Unknown

Eternal life seemed to most Christians – and therefore to most people – in the later nineteenth century to be central to their religion. In April 1875 there appeared an anonymous book, *The Unseen Universe*, which turned out to be written by two eminent physicists, Peter Guthrie Tait of Edinburgh and Balfour Stewart of Manchester. It was one of the successes of the day: there were fourteen editions in thirteen years. Tait, with William Thomson, had published a *Treatise on Natural Philosophy* in 1867 which became a standard advanced textbook because it treated physics from the point of view of conservation of energy – fathering the doctrine, as became two Cambridge men, upon Newton: the 'return to Newton' was for them the key to modernity. Stewart's most important work was on radiant heat; at the Kew Observatory he had collaborated with Warren de la Rue on sunspot investigations. Both Tait and Stewart were staunch churchmen; their book set out to show that modern science gave no reason for abandoning the belief in personal immortality. Like any good piece of natural theology, it also contained a lot of popular science; and many people must have learned about the laws of thermodynamics from reading this introduction to them.

Recent historians, like Owen Chadwick in *The Victorian Church*, are doubtful about the role of science in undermining faith in the latter part of the century.[1] The wealth of Bishops, the problems of the Biblical texts and of other religions, and the morality of both the Old and the New Testaments (arbitrary, and socially unprogressive) were more important factors in the painful losses of faith which Victorians underwent. Despite the increase of unbelief, beginning well before the *Origin of Species* and showing itself in a whole new genre of novels of religious doubt starting with J. A. Froude's *Nemesis of Faith* (1849), the late Victorian period saw the enormous expansion of the missionary enterprise and of church building. For everyone who lost his faith, there must have been at least another who went to outcast London or to darkest Africa to convert the faint and hungry heathen; there was a great deal of Christian confidence, and not simply a Church in retreat before agnostic scientists.[2]

We can see from Royal Institution Discourses again how religion was a part of the scientific culture, along with education and more central topics like energy. In January 1872 Cardinal Manning, the Roman Catholic Archbishop of Westminster, talked about the Daemon of Socrates, concluding that Socrates bequeathed us the 'lesson – that Conscience is the voice of God'. Two years later the Anglican Dean of Westminster, Arthur Stanley, gave a Discourse on what the Catacombs of Rome indicated about early Christian beliefs. Just after Manning had come the agnostic Tyndall, talking about the identity of radiant heat and light; and just before Stanley, the militant anti-christian W. K. Clifford had held forth on the education of the people, and especially on the importance of technical drawing. Audiences might have enjoyed reinterpretations, as when Odling at the same period suggested that phlogiston was an anticipation of the idea of chemical energy, rather than a non-substance. In another reinterpretation, in 1874, the Rev. Sedley Taylor of Cambridge talked of the trial of Galileo, drawing attention to the possibility of a forged Inquisition minute, and the probability therefore that Galileo ought to have been acquitted whatever one thought of the general merits of his theories; his condemnation was the fault of a dirty tricks brigade rather than of the Church.

Nevertheless, however strong the churches remained, there were many sympathetic who believed that they were crumbling, and felt sad, nostalgic or alarmed about it. The late Victorian period was the first in which there were considerable numbers of agnostics, atheists and opponents of organized religion who led lives of exemplary respectability; it could not be urged of Huxley, Clifford or Tyndall that they were irreligious because they were libertines, like the earlier ungodly. Indeed 'atheist' in the seventeenth and eighteenth centuries usually meant somebody who behaved as though there were no moral laws in the world. The great Bishop Joseph Butler of Durham, had begun his *Analogy of Religion* of 1736 with the famous remark:

> It is come, I know not how, to be taken for granted, by many persons, that Christianity is not so much as a subject of inquiry; but that it is, now at length, discovered to be fictitious. And accordingly they treat it, as if, in the present age, this were an agreed point, among all people of discernment; and nothing remained, but to set it up as a principal subject of mirth and ridicule, as it were by way of reprisals, for its having so long interrupted the pleasures of the world.

By the 1870s Christianity had survived and flourished, and it was not now mirth and ridicule from hedonists – the scientific age was a serious one – but assault from intellectuals and moralists which would shake the churches.

In 1874 John Tyndall at the meeting of the British Association in Belfast delivered what came to be known as the Belfast Address. This broke the tradition whereby Presidential Addresses were not expected to be highly personal; the convention was that the President reviewed the scientific developments in the past twelve months, usually appealed for more public support for research and education, and might indicate what avenues seemed most hopeful. Like the Royal Society from its foundation in the 1660s, the BAAS was kept off questions touching party politics or religion, partly in the belief (or hope) that real science could and should be value-free, and partly for pragmatic reasons – an unwillingness to alienate any group. Tyndall chose, and it was brave to do it in Belfast, to mount an attack on organized religion in the name of science. Tyndall was a member of the X-club, a group of prominent agnostic scientists who were also keen on the professionalizing of science; they were an important pressure-group, but by no means representative. To have the spokesman for science take on the churches seemed to many neither congenial nor wise.

Tyndall was a Professor at the Royal Institution, and an accomplished lecturer in that tradition, where arousing interest in a general audience was all-important, in contrast to academics training students, who thought of him as windy and superficial. Darwinism, progress, conservation of energy, the psychology of Herbert Spencer, and physiology wrestling with the nature of life, all led to a world-view very different from that of the churches. Tyndall was not a materialist, but more of a pantheist transferring religious language from God to the world. He saw evolution as 'the manifestation of a Power absolutely inscrutable to the intellect of man. As little in our day as in the days of Job can man by searching find this Power out.' But science was a better way to approach this Mystery than any dogmatic religion could be in the nineteenth century: 'We claim, and we shall wrest from theology, the entire domain of cosmological theory.' For Tyndall, science claimed the unrestricted right to search even on dangerous ground; like Goethe he believed that science ought to be lively, and that commotion was to be preferred to stagnation, the torrent to the swamp. Science must not be an idol; rather it is a pilgrimage, an exercise for the creative minds of every new generation. Tyndall's was scientism with a human face.

The new ideas to which Tyndall referred were also effective within the churches. Though after 1870 all forms of modernism were condemned within the Roman Catholic Church, and St G. Mivart (who had shown sympathies with Darwinism) was excommunicated, other churches and parties came to various degrees of accommodation with contemporary thought. Henry Mansel the philosopher-theologian after all published in 1858 *The Limits of Religious Thought* arguing that we can know almost nothing positive about God; this God is not so far from Tyndall's inscrutable Power. In 1860 the collection *Essays and Reviews* had been published by a group of liberal clergy including Benjamin Jowett, Master of Balliol College, and Frederick Temple, a future Archbishop of Canterbury. Some urged the impossibility of miracles, and the inadequacy of the Old Testament chronology, and all pressed for the treatment of the Bible like any other ancient writing. The so-called Higher Criticism had come late to Britain, and its consequences are still with us in the interpretation of stories of Virgin Birth and Resurrection. Bishop William Colenso, the author of a well-known textbook of arithmetic, proposed a 'kenotic Christology' (in which Jesus had abandoned his divine attributes while on

Earth) in his *St Paul's Epistle to the Romans* in 1861. In the following year Colenso began to publish his *Pentateuch and Book of Joshua*, showing that there were errors of fact there.

Unorthodox clergy could be disciplined, though by the 1860s the process was unedifying to most people and was also liable to overturn by the secular courts. Whereas some sensitive clergy might feel that they could not but take account of recent scientific thinking, men of science who were churchmen would find it even harder to avoid: Faraday's way of keeping a serene faith and active science apart was not so easy in 1870. There had been laymen – Charles Bell, William Prout, John Kidd and Peter Mark Roget – among the Bridgewater authors of the 1830s, and lay theologizing was thus not unknown in Britain. Many scientists were well-read in theological writings, and thus when in 1875 Stewart and Tait produced *The Unseen Universe* they were not doing anything very much out of the ordinary. They sought to show that nothing in modern science required the abandonment of traditional Christianity, and especially the belief in eternal life. Although the book was anonymous in the early editions, the secret was soon out.

Stewart and Tait began with the idea that the great majority of mankind had always believed in some fashion in life after death. By the 1870s there were many who disbelieved and who yet 'retained the nobler attributes of humanity'; but if the bulk of humanity lost its faith then it would be hardly possible to imagine 'civilised and well-ordered communities' surviving. Indeed the authors were appalled at the violence of their society; and hoped (p. 143) that electricity, 'under the direction of skilled physicists and physiologists' might be used for sending 'absolutely indescribable torture (unaccompanied by wound or even bruise) thrilling through every fibre of the frame of such miscreants.' Their hope has been realized, by rather than for miscreants; their rhetoric reminds us that Victorian values did not produce an orderly society. Indeed the hope of eternal life was thought socially important as a sanction, as the authors imply but do not quite state: the prospect of endless rewards and punishments might bring the wicked to their senses. Revivalists in Britain and America dwelt upon the pains of Hell. While the church-dominated governing body of King's College, London, did not object to Lyell's geology, it sacked the theologian F. D. Maurice for suggesting that the indescribable torture of the damned might not go on for ever.

Davy at the beginning of the century had derived comfort (and in personal if not social terms, the idea of eternal life is comforting) from the conservation of matter.[3] He wrote:

> If matter cannot be destroyed,
> The human mind can never die.
> If e'en creative when alloyed
> How sure its immortality.

Stewart and Tait were more aware of the degradation of energy, moral as well as physical, and thus of the impermanence of the world. They stressed the principle of continuity lying behind the order in the universe which has developed through second causes; and they were sympathetic to Darwin. From John Herschel and from Maxwell they took the idea that the atoms of matter, being indistinguishable, are manufactured articles, created by God as alike as shillings from the Mint. Life too never came from inorganic matter, and must have reached the Earth from elsewhere. The principle of continuity was a precondition for science; we must have faith that we shall never be brought to confusion by inexplicable irregularities. The principle was, they believed, shown in the continuity from the solid state through gases to the ether in which light travelled; it connected this seen universe with an unseen one, permanent and underlying this declining world.

Tait and Stewart sought to show not the truth of Christianity, but its compatibility with modern science, and the narrowness of those who contented themselves with what they called 'how' questions and never with 'why'. Agnosticism or scientific materialism was a poor thing which did not take the universe and its order seriously enough. Their book was assailed by Clifford, who asked what the 'sickly dreams of hysterical women and half-starved men' had to do with the 'gentle patience of the investigator that shines through every page of this book'.[4] Clifford saw 'the slender remnant of a system which has made its red mark in history, and still lives to threaten mankind. The grotesque forms of its intellectual belief have survived the discredit of its moral teaching'.

Clifford was, in the imagery of the authors of *The Unseen Universe*, one of those on the voyage who remains in the engine-room seeing how it works, rather than going on deck to see where the ship is going. Maxwell was a more sympathetic critic, seeing the work (perhaps with tongue in cheek) as science baited with 'the allurement of some more human interest'. Stewart and Tait could at best demonstrate the possibility of eternal life; no traveller returned from that bourne with authentic tidings, except Jesus long ago.

The scientific age saw the new phenomenon of the instan-taneous transmission of thought across vast distances. The electric telegraph seemed to those of the mid-nineteenth century to be one of the most amazing feats of science – reasonably enough, if one remembers the long and uncertain times letters took in days of sail. In the 1860s the first cables were laid across the oceans; thanks to Wheatstone and William Thomson, one could communicate with absent friends whom one could no longer see. At the same period spiritualism came from America to promise a similar kind of contact with the dead. In Britain it flourished especially among those for whom orthodox religion had become untenable, providing a promise of immortality apparently much firmer than argument such as that of Stewart and Tait, or of the Bible.[5] Psychic phenomena seemed to cry out for, and lend themselves to, scientific investigation. Francis Galton had investigated the efficacy of prayer by seeing whether members of the Royal Family, often prayed for in church, lived longer than other aristocrats; he found they did not. But this did not really hit orthodox Christianity, for prayers are always subject to the thing asked for being God's will; and He moves in a mysterious way, so that the workings of Providence may not be apparent until long afterwards. Belief in a personal God has little connection with 'evidence'. On the other hand, spiritual-ism was testable, or seemed so: without any background of dogma, it aimed to provide evidence for the survival of loved ones.

The phenomena were very varied, and usually rather exciting. Generally darkness, or weak lighting, was necessary; then tables rocked, objects moved, lights were seen and things sometimes felt, and messages were rapped out. The messages were often banal, but sometimes gave surprising information about a participant which the medium would not have been expected to know. There were affinities with prayer meetings, with professional 'magicians' or 'conjurors', and with scientific entertainment using the mysterious forces of magnetism or electricity. In the seventeenth century some of the first Fellows of the Royal Society had been interested in spirits and ghosts. Joseph Glanvil, for example, believed that if people gave up belief in ghosts and witches they would soon also give up God, and he investigated some supposedly haunted houses. The results are described in his *Saducismus Triumphatus* (1689), aimed at those modern Saducees who deny the survival of the soul.

Unfortunately for Glanvil, many ghost stories are faked; and investigators of the psychic phenomena of two centuries later came up against the same problem.[6] The very name 'conjuror' for an entertainer shows how the mighty sorcerer of the Renaissance is fallen. The history of science is full of priority-disputes, mistakes and dogmatism, as one might expect from a human and provisional activity. As Charles Babbage noted in his *Decline of Science in England* (1830), trimming and cooking happen, but hoaxing and forging, those more serious forms of fraud, are rare. Because science is public knowledge, experiments and observations should be repeatable; but an enterprise that depends on trust will sometimes have advantage taken of it by a rogue.

An area where some phenomena are known to have been faked, where the will to believe among participants may be very strong, and where observation happens under poor conditions, is hardly a promising one for scientists. Some, like Faraday, were highly sceptical but prepared to try an experiment or two. Faraday showed to his satisfaction that the muscles of those pressing upon a table in a seance moved before the table did rather than in response to its motions; he considered further investigation a waste of time. Others, several of them very distinguished, recognized that great care was necessary but believed that they were seeing something so important that it ought to be properly investigated.[7] During the 1870s a group at Cambridge under the leadership of Henry Sidgwick, an eminent philosopher whose *Methods of Ethics* (1874) is a standard work of utilitarianism, began systematic experiments. Some mediums were found to be fraudulent but there seemed to be genuine phenomena underlying the whole business; and it seemed to be the job of earnest doubters to come to grips with it despite its general slipperiness.

In 1882 the Society for Psychical Research was founded; at first it contained both convinced spiritualists and sober enquirers like Sidgwick, but soon the spiritualists left. Sidgwick was the dominant figure in the Society down to his death in 1900. It became cautious, fair-minded and increasingly respectable, though firm conclusions did not emerge from its investigations. In 1886 it published two stout volumes by Edmund Gurney, Frederic Myers and Frank Podmore on *Phantasms of the Living*. The question being investigated was the appearance of a loved one who is far away and at a moment of unexpected crisis, or dying. There were many records of such manifestations, from reliable and intelligent observers, who had often described the curious incident before they could possibly have heard bad news from India or Australia, where at that moment a friend's life was endangered. This research led the authors into experiments on telepathy, which they tried to handle statistically, holding up cards in different rooms, and so on. Just as Darwin tried to build up series of small changes as in dog or pigeon-breeding to enormous evolutionary changes, so they sought continuity between unsurprising telepathy and the seeing of phantasms. To see such hard-headed investigators wrestling with such uncertain data is fascinating. We also get from the book a sense of the terrible waste of young life, even among the professional middle class from which most informants came.

If the results are striking as social and intellectual history – perhaps hardly as science, as a physicist might see it – the book is also striking for the information it gives about the Society for Psychical Research. It was not a group of cranks. In 1886 the President was Balfour Stewart, Tait's collaborator; the Vice-Presidents included the Bishops of Carlisle and of Ripon, A. J. Balfour, the future Prime Minister, Lord Rayleigh, the physicist – who had succeeded Maxwell as the Cavendish professor at Cambridge – as well as Sidgwick and other notables.

The Honorary Members were John Couch Adams, the astronomer; William Crookes, the chemist, of whom we shall hear more; W. E. Gladstone, Prime Minister again early in that year in the midst of the Irish Home Rule struggles; John Ruskin; Lord Tennyson; A. R. Wallace, the co-discoverer of natural selection; and G. F. Watts, the painter. Members of Council were W. F. Barrett, a physics professor from Dublin; Oliver Lodge, then at Liverpool (another physicist); A. Macalister, the Professor of Anatomy at Cambridge; and J. J. Thomson, also of Cambridge and at the beginning of an exceedingly distinguished career in physics. The 'establishment' was fully represented in the Society, which was engaged with problems seen as crucial in late Victorian Britain, and had over 600 members.

Spiritualism had come from America, and had aroused great excitement there. In America the problems of faith in God were becoming acute in a universe of Darwinism and decreasing availability of energy, with materialism perceived as leading to the collapse of civilization. So among the twenty Corresponding Members of the Society we find William James of Harvard, with Edward Pickering, director of the Harvard Observatory, and five other Americans. What is perhaps more surprising is to find five Frenchmen and five Russians among the group, for in these countries one might expect to find that the religious and intellectual perspectives would be rather different. With its careful experiments and its use of statistics the Society was, however, doing work not altogether remote from experimental psychology. The Corresponding Members also included a German, an Italian and, more interestingly, an Indian – Mahádeva Vishnu Káné, BA, of Bombay, whose religious background must have been very different from that of the earnest doubters of Cambridge.

While psychical research thus attracted many eminent members of the scientific community, there were others who like Faraday and W. B. Carpenter were repelled by it. To be seriously concerned with this field was to expose oneself to the possibility of ridicule when it turned out that one had been deceived by a clever confidence trickster; at best it led to controversy, to results which were suggestive rather than conclusive, and away from those straightforward and answerable questions which since

Galileo's time had been the essence of scientific research. Respectable journals which reckoned to cover the whole of science refused to publish psychical research; reports and discussions of it were unwelcome at meetings of scientific societies; and involvement in it had at best an ambiguous effect on a scientist's career.

Very few specialized in this new field; Gurney, Myers and Podmore were among them. To make a profession out of psychical research was hardly possible, even though by the 1880s there were career structures in more-established sciences. Such research had therefore to function as a hobby. We can see the problems if we look at the work of William Crookes.[8] He was unlike Tait or Rayleigh or J. J. Thomson in that he did not have the useful start in scientific life of a Cambridge degree, or indeed any degree. In that, he was like Wallace[9] and like Norman Lockyer, the first editor of *Nature*;[10] but he was unlike them in that he went on to become President of the Royal Society, at the time of the First World War. He belonged to perhaps the last generation in which it was possible to be very successful in science by making one's own way, rather than following a standard pattern. J. J. Thomson wrote of him:

> Crookes was in many ways unlike other English men of science. His well-waxed moustache gave him a somewhat foreign appearance. . . He was a director of public companies, the proprietor and editor of a journal, gas inspector and expert witness, and at one time he owned a gold mine; he did his work in a laboratory of his own . . . He was very quick to observe anything abnormal and set to work to get some explanation . . . In his investigations he was like an explorer in an unknown country, examining everything that seemed of interest.[11]

He was just the person to be drawn into psychical research, and also into work on the boundary of matter and energy, a frontier like that between body and mind.

Crookes was born in 1832 and, having left school at fifteen, he went to the Royal College of Chemistry. There he worked on the unfashionable inorganic chemistry; his science was always to be on the boundary of physics and chemistry. In 1856, after a period in Oxford and at Chester, he returned to London because he was convinced that only there could a career be made in consultancy and journalism. The problem with consultancy was that companies liked to use the eminent, to whom they were prepared to pay high fees; getting started was very difficult. Journalism was more promising to the entrepreneur-scientist, with steam-presses, cheap paper, the abolition of taxes on knowledge, and new readership with expanding education and the growth of specialization. Here again, scientific eminence would be a help; but versatility was a start, and in 1861 Crookes wrote to James Samuelson, a barrister and naturalist involved in publishing:

> I append the headings of a few subjects to which I have given some attention, and which I could write a readable article upon. I have some others, but have mislaid my memoranda. I ought to say, however, that I am in the habit of writing on *any* scientific topic which may arise, and have such excellent opportunities of getting the best scientific information, that I have no doubt I could please you on any subject connected with my department of science which you may suggest.[12]

The subjects were 'Aluminium, Preservation of stone, Divisibility of matter, Electric & lime-light, &c, The infinitely small, in space & time, Spectrum analysis, &c &c.' Aluminium had recently been prepared in bulk by the eminent Parisian Professor Henri Ste-Claire Deville, who became famous as both a teacher and researcher.

In 1861 Crookes made the discovery which brought him the necessary eminence; he identified the new element thallium. This success came through his rapid taking up of the new technique of spectrum analysis of Kirchhoff and Bunsen; he noticed a green line in the spectrum of some impure selenium. After beating off a challenge to his priority by a Frenchman, C. A. Lamy – a discovery like this was a very important bit of property indeed – Crookes was elected into the Royal Society in 1863, his sponsors including Faraday, Tyndall, Odling, Abel, Stewart and Herschel. He spent the next ten years in a classic investigation of thallium and its properties. Unfortunately he found like others that even eminence in science did not automatically bring emolument, and in 1863 he contemplated moving to Manchester; but Samuelson then came up with the proposal of producing a new *Quarterly Journal of Science*, which duly appeared in January 1864 under the editorship of Crookes and Samuelson.

Already in 1860 Crookes had embarked on what turned out to be an important journalistic enterprise, launching the weekly *Chemical News*, which he edited until 1906. This was a complete departure from the pattern of the learned journal in which papers are refereed and polished, and then appear after some months;[13] it brought rapid publication of results, and Crookes also encouraged speculative papers which the staid, and Baconian, Chemical Society would not put into its *Journal*. He reprinted articles from other publications, often in weekly parts; translated papers such as those of Mendeleev; and gave news of industrial developments and of exhibitions; he had also a lively correspondence section. This was particularly valuable at a time when the chemical profession was coming to be seen as having different needs and interests from the learned chemical community. The Chemical Society represented academic chemists, but had not performed any of the professional 'trade union' functions of fixing fees, ensuring that only the duly qualified were allowed to do public analyses, and so on. Crookes operated on both sides of this line; and *Chemical News* is a very valuable guide to what a chemist was in the later nineteenth century.

The *Quarterly Journal of Science* had a different aim. It was to be a serious vehicle of general science. As such, it contained a section of 'original articles', some of them in effect disguised editorials, which were signed, with an indication of status. Then came reviews, really essay-reviews, which were anonymous and discursive, and sometimes formidable, discussing one or more recent books. The resemblance here is to the great nineteenth-century *Reviews*, the *Edinburgh*, the *Quarterly*, the *Westminster*, and others, which did include some scientific works among their criticism but were chiefly concerned with literature, history and politics.[14] By 1873 there were some signed reviews, an adoption of the French practice, which seemed very odd to most Englishmen and never became general, here or elsewhere, in the nineteenth century. We find interesting matter, like the cost of offprints and of binding-cases. The index varies considerably in size from year to year, and must have been trimmed to fit the space available; the number of illustrations fell over time; and the publisher changed. The journal apparently never made money for Crookes after all, but it kept his name before the public; and being an editor does give some power, patronage and influence.

After a row in 1869, Crookes' name disappeared as an editor; but Crookes wrote to Samuelson a 'rather pungent' letter including the passage:

> I am not a child to be frightened by a resolution of a board of directors, or to be soothed by the present of shares (*each carrying £10 liability*) in a company which has been losing money every year from its commencement, and which nothing short of a miracle could make pay a dividend as hitherto managed; neither is my position in Scientific Circles here such as to render my retaining the editorship of any advantage to me – indeed, I am vain enough to think that I confer more than I receive.[15]

As a result, the journal came under Crookes' complete control, and in 1870 he published in it the first of a series of four papers on spiritualism.

Papers on the spirits thus appeared among those on a recent eclipse, on the early days of the Royal College of Chemistry by Hofmann, on patent laws, and on chemical energy. Indeed the general range was very similar to what we have met among Royal Institution Discourses at the same date. But spiritualism remained beyond the pale, like phrenology and unlike orthodox religion, despite the claims of some that it was really the same as primitive Christianity. Journals which published it, and editors who included it, ceased to be fully respectable; this message got through to Crookes, and he stopped his active propaganda, though apparently never ceasing to believe that some of the phenomena he had witnessed, often in distinguished and reliable company, were genuine.

In 1879 Crookes sold the *Quarterly Journal*; it lived on until 1885, becoming in its last years a monthly – but it had never been a leading journal, and general scientific reviewing never caught on. The great success among new journals of the period was *Nature*, beginning in 1869 and, like *Chemical News*, coming out in newspaper format each week. Unable to investigate further the boundary between body and spirit, Crookes moved on to work on that mysterious borderland, as he called it, of matter and energy. He invented the radiometer, a perplexing scientific toy in which a small propeller contained in an evacuated bulb has its vanes blackened on one side; when light falls on it, the propeller turns. Crookes could not explain satisfactorily what was happening, but Stokes and others involved with the dynamical theory of gases were ultimately able to account for it in terms of collisions and energy.

Crookes took up Faraday's work on the passage of electricity through gases, and was excited by a passage in an early lecture of Faraday's where he had speculated that there might be a 'fourth state of matter' simpler than the gaseous, just as gases are simpler than liquids; they all expand alike when heated, for example. As pumps improved, Crookes got lower pressures than Faraday had been able to do;[16] and eventually he discovered cathode rays, which he interpreted as this fourth state. He also continued with spectrum analysis,[17] mapping spectra and struggling to work out which lines belonged to which elements among the confusing Rare Earth Metals like Lanthanum. He found that glass containing these metals was useful for sun-glasses. The analytical techniques of which he was a master were inadequate to separate these metals completely, and he concluded that they merged into one another, being varieties rather than true species produced by inorganic evolution. Crookes' lectures were illustrated by brilliant and striking experiments, especially with the cathode rays, which he believed to be a stream of particles; but his physics and mathematics were not sufficient for him to produce quantitative work in this field. He also had, as J. J. Thomson remarked, the temperament of the explorer rather than the cool, logical mind of one determined to test a definite theory.

Crookes was thus very important in preparing the way for the 'discovery' of the electron by J. J. Thomson in 1897, in an experiment in which cathode rays were deflected by a magnetic field, and then brought back to the zero point by an electric field. This set-up allowed Thomson to compute the ratio of charge to mass for the particles which the experiment seemed to show must compose the rays. He found that the corpuscles, or electrons as they were later called, must be much smaller than those which Crookes had envisaged; and Crookes later came round to his view. To discover the electron is rather different from discovering the kangaroo, because it is a theoretical entity inferred from certain abstruse experiments; and Thomson's apparently crucial experiment turned out in the twentieth century not to be so.

In Germany, where Hertz had discovered the radio waves in accordance with Maxwell's theory of the electromagnetic field, and where in 1895–7 W. K. Röntgen (using a Crookes tube to generate cathode rays) had discovered X-rays,[18] the general belief was that the cathode rays would also turn out to be a form of wave in the ether. In the twentieth century, experiments devised to test this idea did indeed find that a beam of electrons could be diffracted like light; the electron microscope depends on this analogy. But in 1900 these uncertainties were still in the future; there were clouds, as William Thomson described them, hanging over some classical theories, but as yet no thunder and lightning.

The success of Crookes and his successors with evacuated tubes meant that the passage of electricity through gases was more studied and better understood than that through solids. The solid state seemed more complicated; and because of the problems of purity, which had plagued Crookes in his work on lanthanum and its congeners, it was very difficult to get consistent results with interesting elements like selenium. Crookes' biographer Fournier d'Albe worked and wrote on this, which had baffled various investigators because its resistance changed under exposure to light – but inconsistently if there is impurity. It was for these reasons that the first 'valves' for radios and televisions and computers were evacuated glass vessels, descendants of Crookes' radiometer and cathode ray tube, rather than semiconductors.

Uncertainty and the reinterpretation of crucial pieces of work were not yet necessary in physics in the 1870s and 1880s, as they seemed to many to be in the human sciences and in the sphere of religion under the pressure of evolutionary theory and the Second Law of Thermodynamics. Evolutionary explanations of society became the norm,[19] especially through the writings of E. B. Tyler such as *Primitive Culture* (1871); while the excessive breeding of the unsuitable worried eugenicists such as Francis Galton, Darwin's cousin and the inventor of fingerprinting.[20] Among discussions of religion, *The Unseen Universe* was unusual for its emphases on the conservation of energy and the running down of the universe; usually it was the idea of development which was more prominent. This need not involve rejection of traditional Christianity; Newman had written his famous essay on the development of doctrine in 1845, long before the *Origin of Species*; and the idea was reworked in the Darwinian light of 1889 in *Lux Mundi*, a collection of articles edited by the prominent high-churchman Charles Gore.

Nevertheless, it seems probable that the scientific spirit, at least as seen by non-scientists, encouraged more critical and liberal attitudes to religion; this can be seen in the Hibbert Lectures, which began in 1878. Robert Hibbert had died in 1849 leaving money for religious purposes, which was at first applied to theological education. The bequest was for 'the spread of Christianity in its most simple and intelligible form, and to the unfettered exercise of private judgement in matters of religion.' A group of eminent liberal divines petitioned the trustees to endow a series of lectures on unsettled problems in theology. They referred to the 'traditional restraint, from which all other branches of inquiry have long been emancipated': ecclesiastical interests and party predilections, which meant that theology 'fails to receive the intellectual respect and confidence which are readily accorded to learning and research in any other field'. There was, they believed, no reason 'why competent knowledge and critical skill, if encouraged to exercise themselves in the disinterested pursuit of truth, should be less fruitful in religious than in social and physical ideas'.

The first lecture series was given by Max Müller, a signatory, in the spring of 1878; the venue, the Chapterhouse at Westminster Abbey, was arranged by another signatory, Dean Stanley; the course proved so popular that Müller gave each lecture twice. He spoke of Indian religion; and indeed comparative religion was a subject becoming popular, a region in which scientific study seemed possible. Its alarming feature to the orthodox was that it might make Christianity seem simply one faith among others, rather than simply truth. Müller's lectures were followed in 1879 by those of Peter Renouf, originally from Guernsey, an associate of Newman, an Inspector of Schools and an Egyptologist; and then in 1880 by Ernest Renan's, coming nearer to home with an investigation of the influence of Rome on Christianity and the development of the Catholic Church.

By the end of the nineteenth century, then, there was a strong feeling that religion, like everything else, should be studied in the light of science. In all fields, there was still much that was unknown: in Herbert Spencer's philosophy, and in Tyndall's Belfast address, there was room for the unknowable; but it seemed reasonable to hope that under pressure from inquiring minds this would shrink, if not ultimately wither away altogether. Just as Crookes' experiments on cathode rays not merely revealed but solved some mysteries, so science would gain ground steadily upon the dark, reducing the area of the unknown. The end of the nineteenth century was the climax of the Age of Science; what remains for us is to glimpse briefly the aftermath of that confident period in which it seemed that all the interesting questions must have answers, and that it was the business of natural philosophy to find them out.

12

Science Comes of Age

By the end of the nineteenth century science had become a dominant force in intellectual and practical life. In Japan it had even secured a missionary success as great as that of the Christian churches in Africa.[1] The different ways in which people lived in Europe and North America in 1800 and in 1900 had much to do with applied science. Its effects also spread much wider, with new trades demanding raw materials from all over the world while new industries undermined old activities, like indigo-growing in India. Attitudes to God and man were modified as science and 'Victorian Values' interacted, in slightly different ways in different places. What is probably most interesting in past science is what is characteristic, rather than what we suppose 'true'; in the nineteenth century what science, or that 'scientific method' which might be applied far beyond the boundaries of a university faculty of science, was supposed to provide was certitude. There were a few areas in which it could do so, but even chemical analyses could turn out wrong. Thus the presence of helium in certain rocks was missed because those analysing them thought that this unreactive gas must be nitrogen, in accordance with the practice of the day. In more complex questions of health and diet the scope for erroneous dogmatism was, and is, much greater.

Men of science themselves recognized by the end of the century that they were not dealing in certainties. Tyndall with his emphasis on creativity and imagination in science, and T. H. Huxley with his on 'working hypotheses', appreciated that there were areas where high probability was the best that could be hoped for. If indeed the criteria in science are pragmatism and coherence, then 'truths' are provisional and will change with our notions of what we want from it, and of what it must cohere with. Different theories work in different social and intellectual contexts, and make different observations and experiments interesting. Maxwell in his statistical interpretation of the Second Law of Thermodynamics had gone deeper. He seemed to have shown that physics itself rested upon uncertainty, and that predictions and explanations are based on probability – overwhelming in most ordinary cases.

Despite such doubts in philosophical moments, scientists were generally happy that they had got the right end of the stick; that their beliefs about the world were truer than those of their ancestors. Certain crucial experiments had demonstrated some propositions, putting them, it seemed, beyond doubt. The work especially of Young, Fresnel, Arago, Fizeau and Foucault had proved that light was a wave motion. Interference fringes; the bright spot in the centre of the shadow of a sphere; light going faster in air than in water – all were predicted by the wave theory and all verified. The idea of transverse vibrations, like the waves of the sea, could account for polarization and for the passage of light through calcite crystals where two images are seen. Every test indicated the power of this theory; and then Maxwell was able to unify optics with electricity and magnetism. One clear example of scientific progress was the abandonment of the 'exploded' Newtonian view that light was corpuscular; and belief in the ether and its waves seemed to Tyndall, for example, to be thoroughly rational and sound, in contrast to some other beliefs.[2]

There were indeed some clouds over this sunny scene. To account for some curious features of radiation, Max Planck of Berlin University had in 1897–1901 introduced the idea that energy might be transmitted in packets, or quanta, of definite size. In 1904 Einstein,[3] then still unknown, accounted for the photo-electric effect, where a current is generated by light falling on for example potassium, in terms of quantum theory: light behaves as though it were a stream of corpuscles, those at the blue end of the spectrum carrying more energy. To pepper potassium with the small-shot of red light produces no effect, whereas even a stray bullet of blue light knocks out electrons and sets up a current. Here were crucial experiments establishing incompatible theories of light. In the following decade electrons were diffracted, like light waves, although J. J. Thomson's experiment seemed in 1897 to have demonstrated that they were particles.[4] For the twentieth century, fundamental science seems closer to the Platonic picture of likely stories than to the Baconian one of established generalizations. It is more like other human activities.

At the same time the ether – invoked because if light is a wave motion, there must be something to wave – came into disrepute. There had been in the nineteenth century all sorts of questioning about the nature of the ether,[5] which seemed to have to be a very strange sort of stuff; and about the Earth's motion, or lack of it, relative to the ether. If the Earth moved through it like a ship at sea, then waves sent forwards would travel faster than those sent backwards; but no such differences were detectable, and Einstein proposed that the ether be dropped, and the velocity of light taken as a constant of nature. Thus began an exciting period in physics, in which the certainties of the classical period (our Age of Science) were lost, but in which the science became thereby much more lively and interesting.

The intellectual certainties of the scientific age thus became eroded in the opening years of our century,[6] and science ever since has been seen as a more provisional activity than it was; at any rate by those reflective about it. Chemists speak confidently about structures of very complex molecules like DNA, but the fate of the wave theory and of conservation of energy (energy now seen as convertible into matter following Einstein's equation, $E=mc^2$) mean that one can no longer take for granted even well-established theories in mature sciences. It was not just a sign of the backwardness of chemists that they all believed in phlogiston until Lavoisier put them right: such revolutions can happen from time to time in the most prestigious sciences. The idea that science might be completed soon has also receded, in favour of the view of Chalmers in the early nineteenth century, that the greater the circle of light, the greater the circumference of darkness.

The moral certainties lasted a little longer, for science still seemed an innocent activity. But the Royal Society over which Crookes presided during the First World War was very different from that directed by Banks in the French wars which had ended after Waterloo a century earlier. Then science had seemed above the conflict, whereas by the twentieth century it was well in it. The Haber process for making synthetic ammonia promised cheap fertilizers, but until the end of the war it was used only for munitions. The Second World War was even more a scientific one; and it was with the explosion of the atomic bomb that science came of age, had to cover its nakedness, and was driven from the garden of innocence. Like every other human activity, it could be employed for good or bad ends; for penicillin and for poison gas. Knowledge had indeed proved to be power, as Bacon said it would.

Davy has been called the apostle of applied science, but the Age of Science in that sense only really began about the middle of the nineteenth century with new industries – chemical and electrical – which were based directly on scientific work. They have been followed by all sorts of others, so that ours is an age of applied science in a way in which the nineteenth century was not. The promises of useful knowledge made by Bacon and his successors were redeemed some two and a half centuries later; and it was chiefly from this time, about the 1870s, that governments began to support science seriously, rather than intermittently for cultural prestige.

Nevertheless it was in that century – which Leslie Stephen, the eminent Victorian man of letters, said nobody would ever want to revive – that science really became an important part of culture. It was more than 'organized common sense': that had indeed lain behind early industry, but not all that is rational is scientific, at least in our modern sense. What science seems to require is satisfactory explanation; and those who are anti-scientific are opposed to certain kinds of explanation, which they usually deem narrow and reductive. There is method even in madness, and it would be very odd to try to live irrationally. Edward Lear's Old Man of the Hague, who bought a balloon to examine the Moon, indeed had ideas which were excessively vague, but not irrational. What would be irrational would be to wish passionately to examine the Moon, to believe that a balloon was the ideal vehicle for this, to be able to buy one, and not do it; perhaps buy a submarine instead.

We see this most clearly in medicine. Homoeopathy and

acupuncture and other 'alternative' therapies have their rationale, though sometimes popular medicine may be excessively vague – as in *Herbals*, where many plants seem to cure most diseases. The point is that they are condemned as quackery by scientific medical men because they lack theory; they cannot give satisfatory explanations, in accordance with current world-views, logical theory and fashion. Fashion may seem too frivolous a thing to be introduced into the realm of the intellect, and especially of science: but our story of the Age of Science has been one of a succession of fashions, among other things.[7] In France, in Germany, in natural theology, in political economy, in chemistry, in evolution, in thermodynamics, people sought for and found satisfactory models. What is interesting in the history of science is the change over time in the demands made of scientific explanation. These may well be improvements, in that the newer theories embrace more phenomena and connect better with those in other sciences; but there will always be some loss, some observations that become beneath notice. Simple falsifications are very rare, and confirmations impossible in a strict sense.

At the beginning of the century we saw the prevalence of national styles, with very different institutions and expectations in different countries of Europe. By the end of the century these had been considerably ironed out, partly because communications were much easier, partly because the sciences were much more organized bodies of knowledge and partly because science was much more cosmopolitan. For most of the century, as for preceding centuries, science had been the business essentially of European men. They were the stars, though there might be women and non-Europeans among the supporting cast. Women played important parts as translators, as authors of popular works, as illustrators and colourists, and increasingly in astronomical observations; but the openings available to them were much restricted at all levels. Similarly, Asians, Africans and Amerindians were necessary as guides for scientific explorers and collectors of natural history, sometimes as scientific travellers themselves (notably in the Himalayas), and as illustrators; that is, as assistants.[8] There were strong currents of what we would call racism and sexism among men of science in the later nineteenth century,[9] which went easily with a kind of social Darwinism.[10] They supposed that brain capacity went with mental capacity, thus putting European males at the top of the scale; and they feared degeneration, the waning of that spirit that won empires.[11]

Against this can be set the new cosmopolitanism. By the end of the century the USA, Russia and Japan were emerging as very important centres of science, the first two having been essentially provincial down to the second half of the century, while Japan kept aloof from international contacts of all kinds down to that time. It may be significant that when in 1908 Arthur Schuster gave a course of lectures on The Progress of Physics he delivered them at the University of Calcutta. Schuster had worked with Maxwell, and at Manchester in 1875–6 had delivered what he believed was the first systematic course based on Maxwell's electromagnetic theory to an audience of three, including the young J. J. Thomson. His reminiscences of graduate work at Cambridge and in Germany make fascinating reading; as does his unhappiness with the new physics of models and uncertainty. By 1908 he could get a much larger audience than three in India, for an admittedly non-technical lecture-course, but one about on the level of Royal Institution Discourses. By the end of our Age of Science, the sciences had lost their close link with any one national culture; with it went the easy links through metaphor to literature and back again which had characterized earlier years.

Half a century before, the eminent theologian J. B. Mozley had written that Christianity could wield the weapon of science in India.[12] He argued, in an essay first published in the same year as the *Origin of Species*, that Christianity must be compatible with new science, whereas false religions, like that of the Hindus, would not be. It did not work out quite as simply as that; but as well as showing ethnocentrism from our point of view, Mozley's remark can remind us that the Age of Science was also an age of religion. The point is that in the nineteenth century, science was entwined with other activities; they were not simply counter-weights or antagonists. By the twentieth century things came to look rather different, as the sciences came of age and became less connected with local cultures. Metaphors from information science are less powerful than 'the struggle for existence'.

The publicists for science in the early days, like Francis Bacon, had urged that it was not merely an intellectual activity but also a practical and a social one. Understanding was not the whole object: in competition or in conjunction with other disciplines, science was going to transform lives, and bring power; and it was also something not to be done by solitary geniuses, but a cumulative enterprise using teams of people of different talents. These last aspects, bringing with them government support and a career structure, became real in the Age of Science. We may hope that in this case Mozley was wrong when he wrote that 'the course of things hands over institutions from genuine to hollow hands';[13] but we may feel that a golden age of science, an age of innocence, is past.

It would be perverse to deny that there is progress in the sciences; but it is not simple progress towards some millenium. The circle widens, and the solution to each problem raises in its turn more problems. There seems to be no goal, and no way of knowing if anything in present science will turn out in the long run to be true. Probably it may go the way of the wave theory of light, and be no more than a model useful in some situations. The Age of Science represented a time of greater confidence than ours, with different scepticisms and credulities. This makes it fascinating to study. Science, as a discipline, a practice, and an institution, was a vital feature of the nineteenth-century world, and thus of many unexpected features of our world; and if we remember that the sciences were not born yesterday, we shall enjoy them more and evaluate them better. If our world is to survive, all our skills, and not just scientific ones, are demanded; and knowing about science (as well as knowing some science) is as important as knowing other kinds of history. The great problem which the Age of Science has bequeathed to us sorcerer's apprentices is perhaps that of specialization, which solved problems our ancestors inherited, but leaves us alienated. There is no easy solution, but seeing how we got here is a help.

Further Reading

The historian of science does not fit easily into any specialized department, which is partly why the subject is so interesting and genuinely interdisciplinary. The first practitioners were generally scientists, active or retired, who were anxious to see how we had got to where we are now. Like lawyers approaching constitutional history, they were interested in useful precedents; they expected to find progress; and they looked particularly for precursors of modern views. This kind of history is sometimes called Whig history, after those like Macaulay who saw the whole of history moving towards Victorian liberalism; it gives a pattern, but it does not lead to sympathy with the losers, and the heroes may have to change as a wave theory of light gives way, for example, to one involving particles. Relations between professional historians of science and scientists interested in the history of their discipline are uneasy but often fruitful.

Twentieth-century historians are usually less willing to see progress in history; and historians of science have tried to follow them in seeing the science of a period as typical of the day rather than as more or less close to our science. This involves placing past science carefully in context; looking at institutions, not in the spirit of celebrating a centenary but of critical analysis; being chary of moral judgements; and remembering that the past is both like and unlike the present.

During and since the Age of Science, philosophers have been much interested by the ideas involved in it, and by the prospect of devising a method of arriving at truth. The difficulty is that history is usually messy; that scientists are humans rather than desiccated calculating machines; and that to fit the frameworks devised by philosophers, history must be rationally reconstructed. There are fascinating case studies from the past which illuminate the methods and the kind of explanations used in the sciences; and the History of Ideas (which overlaps with the History of Science) is a difficult but rewarding activity on the boundaries of history and philosophy.

Science is an activity which has strong social connections, and sociological ideas have also proved fruitful for the historian of science. They can lead to history of science with the science left out: to analyses of societies which happened to practice science but might have been growing leeks, because it is the status of the members rather than the nature of their activity which interests the researcher. There are undoubtedly parallels between scientists and other social groups; the analyses of provincial scientific societies, of bodies like the British Association, and of elite groups like the Royal Society and other Academies, can tell us a great deal about scientists. They are after all people with careers to make, longing for recognition, competing for status, as well as searching for truth; and it has been suggested that science has functioned as an opiate for the people and an instrument for social control.

The historian of science thus finds himself making eyes at scientists, at historians, at philosophers, at sociologists, and also at historians of art, medicine, literature and technology; nobody can please them all, and all have useful contributions to make to the history of science. The aim of this book is to indicate how accessible different aspects of past science are to those with different interests; and how both our science and our past can be understood only when the Age of Science is studied.

In the reading lists which follow, I have tried to pay my intellectual debts in some degree; and also to suggest new works which bear in various ways on the questions discussed in the chapters to which they refer. Some of the references are to books – some to be referred to, and some read – and others are to journals; those who get really interested in the field will want to join a society arranging lectures, discussions and publications in the history of science. It is sufficiently far from a narrow specialism for many authors to make the effort to make their books and papers readable; and browsing from the reading-lists should be a pleasure rather than a discipline. To enjoy history, and even to think historical explanation the fundamental kind, is not to live in the past, but to admit that loss of memory is no fun.

In the following lists and Notes, *BJHS* refers to the *British Journal for the History of Science*.

1 The Age of Science?

Standard works on earlier science are A. R. Hall, *The Revolution in Science* (London, 1983); K. Thomas, *Man and the Natural World* (London, 1983); R. Porter, *The Making of Geology* (Cambridge, 1977). For the Age of Science, the classic work is J.T. Merz, *A History of European Thought in the Nineteenth Century*, 4 vols (Edinburgh, 1904–12); later work is largely a series of footnotes to Merz, but naturally many of his perspectives differ from ours. Splendid illustrations are in L. P. Williams, *Album of Science* (New York, 1978). A recent general book is C.A. Russell, *Science and Social Change, 1700–1900* (London, 1984); this, like S. F. Cannon, *Science in Culture* (New York, 1978), includes social and intellectual history, while D. Mackenzie, *Statistics in Britain, 1865–1930* (Edinburgh, 1981), seeks ultimately social explanations. For science as a social, practical and intellectual activity, see my *The Nature of Science* (London, 1976); and my 'The history of science in Britain: a personal view', *Zeitschrift für allgemeine Wissenschaftstheorie*, 15 (1984) 343–53. On scientific thinking, see B. Barnes, *About Science* (Oxford, 1985).

Essential reference works include C. C. Gillispie (ed.), *Dictionary of Scientific Biography* (New York, 1970—; T. I. Williams (ed.), *A Biographical Dictionary of Scientists* (London, 1969); W. F. Bynum, E. J. Browne and R. Porter (eds), *Dictionary of the History of Science* (London, 1981) – conceptual, not biographical; and J. Wintle, *Makers of Nineteenth-century Culture* (London, 1982); as well as standard national biographical dictionaries.

There are various societies devoted to the history of science or to aspects of it. These include the British Society for the History of Science, which publishes its *Journal* (*BJHS* for short) and holds meetings and conferences; the History of Science Society in the USA, which publishes *Isis*; the Society for the History of Alchemy and Chemistry, publishing *Ambix*; and the Society for the History of Natural History, publishing *Archives* and various special publications. The Newcomen Society publishes *Transactions* on the history of technology; and there are also journals not published by societies, including *History of Science*, *History of Technology* and *Annals of Science*.

2 Rivalry with the French

Standard works are M. P. Crosland, *The Society of Arcueil* (London, 1967), and *Gay-Lussac* (Cambridge, 1978). On Lamarck and his milieu, see R. W. Burckhardt, *The Spirit of System* (Cambridge, Mass., 1977) and L. J. Jordanova, *Lamarck* (Oxford, 1984). S. Carnot, *Réflexions*, ed. R. Fox (Paris, 1978); an English version is in preparation. C. C. Gillispie, *The Montgolfier Brothers* (Princeton, 1983). L. J. Klosterman, 'A research school of chemistry in the nineteenth century', *Annals of Science*, 42 (1985) 1–40 describes Dumas' laboratory. P. Lervig, 'Sadi Carnot and the steam engine', *BJHS*, 18 (1985), 147–96; R. Fox, 'Science, industry and the social order in Mulhouse, 1798–1871', *BJHS*, 17 (1984), 127–68. On Poisson's election, R. W. Home, 'Poisson's memoirs on electricity', *BJHS*, 16 (1983), 239–59; A. Petit, 'L'esprit de la science anglaise et les Français au XIXème siecle', *BJHS*, 17 (1984), 273–93 describes the reverse flow, from England to France. D. Outram (ed.), *The Letters of Georges Cuvier: a summary calendar* (Chalfont St Giles: BSHS), 1980 and D. Outram, *Georges Cuvier* (Manchester, 1984). On Davy, see S. Forgan (ed.), *Science and the Sons of Genius* (London, 1980); on medicine, A. D. C. Simpson (ed.), *Joseph Black* (Edinburgh, 1982); M. Foucault, *The Birth of the Clinic*, trans. A. M. Sheridan Smith (London, 1973); J. Woodward, *To Do the Sick No Harm* (London, 1974). G. Vevers, *London's Zoo* (London, 1976); and on Kew, W. Blunt, *In for a Penny* (London, 1978). On Faraday, see D. Gooding, and F. James *Faraday Rediscovered* (London, 1985); and on Davy at the Royal Society, D. P. Miller, 'Between hostile camps', *BJHS*, 16 (1983) 1–47. N. Baudin, *Journal*, trans. C. Cornell (Adelaide, 1974); and on Flinders' voyage, the life of Robert Brown, D. J. Mabberly,

Jupiter Botanicus (Braunschweig, 1985); on voyages generally, M. Deacon, *Scientists and the Sea* (London, 1971). D. King-Hele, 'Erasmus Darwin', *Interdisciplinary Science Reviews*, 10 (1985) 170–91.

3 Wrestling with God

C. A. Russell, *Cross-currents* (London, 1985), emphasises evangelical, Biblical, Christianity more than I do; on high-churchmen, see P. Butler (ed.), *Pusey Rediscovered* (London, 1983). On eighteenth-century materialism, see J. W. Yolton, *Thinking Matter* (Oxford, 1984) and the essay-review by G. Cantor, *History of Science*, 23 (1985) 201–6; for later, my *Transcendental Part of Chemistry* (Folkestone, 1978). A. W. Colver and J. V. Price (eds.), *Hume on Religion* (Oxford, 1976); on Coleridge's science, T. H. Levere, *Poetry Realized in Nature* (Cambridge, 1981); on Whewell, R. Yeo, 'William Whewell', *Annals of Science*, 36 (1979), 493–516 and 'Baconianism', *History of Science*, 23 (1985) 251–98; on *Vestiges*, R. Yeo, 'Science and intellectual authority', *Victorian Studies*, 28 (1984), 5–31. On Buckland, N. A. Rupke, *The Great Chain of History* (Oxford, 1983). J. H. Brooke is writing a book on natural theology; see also the relevant chapter in my bibliographical study *Natural Science Books in English, 1600–1900* (London, 1972). M. Hunter, *Science and Society in Restoration England* (Cambridge, 1981); L. J. Jordanova and R. Porter (eds), *Images of the Earth* (Chalfont St Giles (BSHS), 1978). J. H. Newman, *A Grammar of Assent*, intr. N. Lash (Notre Dame, 1979). On William Prout, chemist and natural theologian, see W. H. Brock, *From Protyle to Proton* (Bristol, 1985). On Swainson, see my papers forthcoming in *Archives of Natural History*, and in J. North and J. Roche (eds.), *The Light of Nature* (Dordrecht, 1985), pp. 83–94.

4 The German Challenge

On Newton and Leibniz, A. R. Hall, *Philosophers at War* (Cambridge, 1980). On teleological biology, T. Lenoir, *The Strategy of Life* (Dordrecht, 1982). S. Hahnemann, *Organon of*

Medicine, trans. J. Künzli, A. Nandi and P. Pendleton (London, 1983); on eminent Germans, see R. Paul, 'German academic science and the mandarin ethos', *BJHS*, 17 (1984), 1–29, and E. Du Bois-Reymond, P. Diepgen and P. C. Cranefield (eds.), *Two Great Scientists*, trans. S. Lichtner-Ayed (Baltimore, 1982) (Ludwig/Du Bois-Reymond correspondence); W. H. Brock, 'Liebigiana', *History of Science*, 19 (1981) 201–18. K. Hufbaur, *The Formation of the German Chemical Community, 1720–1795* (Berkeley, 1982) is a valuable run-up to our period, while for the later history of the 'German' science of physiology in Britain and in the USA, see G. L. Geison, *Michael Foster and the Cambridge School of Physiology* (Princeton, 1978) and R. E. Kohler, *From Medical Chemistry to Biochemistry* (Cambridge, 1982). On Richard Taylor, W. H. Brock and A. J. Meadows, *The Lamp of Learning* (London, 1984); on the 'simultaneous discovery' of energy conservation, T. S. Kuhn, *The Essential Tension* (Chicago, 1977); on German universities, P. R. Sweet, *Wilhelm von Humboldt* (Columbus, 1978–80). On science in North America, N. Reingold (ed.), *The Papers of Joseph Henry* (Washington, 1972–); T. H. Levere and R. Jarrell, *A Curious Field-book* (Toronto, 1974); S. G. Kohlstedt, *The Formation of the American Scientific Community* (Urbana, 1976); A. P. Molella, 'At the edge of science', *Annals of Science*, 41 (1984), 445–61; L. Owens, 'Pure and sound government', *Isis*, 76 (1985), 182–94. On Hegel and science, see Hegel's *Philosophy of Nature*, ed. and trans. M. J. Petry (London, 1970); a conference on this theme was held in 1985 at Amersfoort, and the proceedings will be published under the editorship of Petry and R. Horstmann, Stuttgart, forthcoming. On fashion, see E. Wilson, *Adorned in Dreams* (London, 1985).

5 Arguing with Sceptics

On phrenology, see the papers by G. N. Cantor and by S. Shapin 'on Edinburgh phrenology' in *Annals of Science*, 32 (1975), 195–256; and by M. Shortland, forthcoming in *BJHS*. On plurality of worlds, J. H. Brooke, 'Natural theology and the plurality of worlds', *Annals of Science*, 34 (1977), 221–86;

S. J. Dick, *Plurality of Worlds* (Cambridge, 1982). G. Tytler, *Physiognomy in the European Novel* (Princeton, 1982); and see S. Shuttleworth, *George Eliot and Nineteenth-century Science* (Cambridge, 1984); G. Beer, *Darwin's Plots* (London, 1983); R. O'Hanlon, *Joseph Conrad and Charles Darwin* (Edinburgh, 1984), on interrelations of science and literature. M. Pollock (ed.), *Common Denominators in Art and Science* (Aberdeen, 1983), covers a range of arts and sciences. On theories of light, and on matter theory see G. Cantor, *Optics after Newton* (Manchester, 1983), and my *Atoms and Elements*, 2nd edn. (London, 1970); F. Gregory, 'Romantic Kantianism and the end of the Newtonian dream in chemistry', *Archives internationales d'histoire des sciences*, 34 (1984), 108–23; C. Smith, 'Natural philosophy and thermodynamics', *BJHS*, 9 (1976), 293–319; T. Kuhn, *Black Body Radiation and the Quantum Discontinuity* (Oxford, 1978). On *Vestiges*, M. B. Ogilvie, 'Robert Chambers and the nebular hypothesis', *BJHS*, 8 (1975), 214–32; J. H. Brooke, 'Richard Owen, William Whewell, and the *Vestiges*', *BJHS*, 10 (1977) 132–45. A. D. Morrison-Low and J. R. R. Christie (eds), *Martyr of Science*: *David Brewster* (Edinburgh, 1984).

6   Debate about Animals

See my *Ordering the World* (London, 1981); J. Browne, *The Secular Ark* (New Haven, 1983); M. P. Winsor, *Starfish, Jellyfish, and the Order of Life* (New Haven, 1976); D. E. Allen, *The Naturalist in Britain* (London, 1976); P. R. Sloan, 'Buffon, German biology, and the historical interpretation of biological species', *BJHS*, 12 (1979) 109–53; A. Desmond, *Archetypes and Ancestors* (London, 1982), and 'Richard Owen's reaction to transmutationism in the 1830s', *BJHS*, 18 (1985), 25–50. I. S. Majneep and R. Bulmer, *Birds of my Kalam Country* (Oxford, 1977) is an example of ethno-taxonomy; M. S. Merian, *The Wonderful Transformations of Caterpillars*, ed. W. T. Stearn (London, 1978); W. T. Stearn, *Botanical Latin*, 2nd edn. (Newton Abbot, 1973). C. Lyell, *Scientific Journals on the Species Question*, ed. L. G. Wilson (New Haven, 1970); D. H. Barrett,

D. J. Weinshank and T. T. Gottleber, *A Concordance to Darwin's Origin of Species* (Ithaca, 1981); C. Darwin, *Correspondence*, ed. F. Burkhardt, S. Smith *et al.* (Cambridge, 1985—); J. R. Moore, *The Post-Darwinian Controversies* (Cambridge, 1979); D. R. Oldroyd, *Darwinian Impacts* (Milton Keynes, 1980); P. J. Bowler, *The Eclipse of Darwinism* (Baltimore, 1983); and on Darwinian scientism, J. C. Greene, *Science, Ideology and World-view* (Berkeley, 1981); J. R. Durant, 'Scientific naturalism and social reform in the thought of A. R. Wallace', *BJHS*, 12 (1979), 31–58; M. Ruse, *Taking Darwin Seriously* (Oxford, 1985). On eugenics, see the papers by L. A. Farrell, D. MacKenzie, R. Love, and J. R. Searle in *Annals of Science*, 36 (1979), 111–69. On careers in natural history, see S. Sheets-Pyenson, 'Geological communication', *Bulletin of the British Museum (Nat. Hist.), Hist. Series*, 10 (1982), 179–226; J. Secord, 'John W. Salter', in A. Wheeler and J. H. Price (eds), *From Linnaeus to Darwin* (London, 1985), pp. 61–75. On museums, see D. Ripley, *The Sacred Grove* (London, 1970); R. Desmond, *The India Museum* (London, 1982); A. E. Gunther, *A Century of Zoology at the British Museum* (London, 1975); M. Girouard, *Alfred Waterhouse and the Natural History Museum* (London, 1981).

7 Discourse in Pictures

See R. B. Freeman, *British Natural History Books* (Folkestone, 1980); my *Zoological Illustration* (Folkestone, 1977); A. Ellenius (ed.), *The Natural Sciences and the Arts, Acta Universitatis Upsaliensis*. 22 (Uppsala, 1985); M. J. S. Rudwick, 'A visual language for geology', *History of Science*, 14 (1976), 149–95; S. Hyman, *Edward Lear's Birds* (London, 1980); W. T. Stearn (ed.), *The Australian Flower Paintings of Ferdinand Bauer* (London, 1976); G. C. Saur, *John Gould* (Melbourne, 1982); T. Keulemans and J. Coldeway, *Feathers to Brush* (Melbourne, 1982). B. Bracegirdle, *A History of Microtechnique* (Ithaca, 1978). Useful material on printing and publishing is to be found in M. H. Black, *Cambridge University Press, 1584–1984* (Cambridge, 1984). Books using old illustrations include

A. Rutgers, *Birds of New Guinea*, illus. J. Gould (London, 1970); G. E. Lodge, *Unpublished Bird Paintings*, text by C. A. Fleming (London, 1983); C. G. Finch-Davies, *The Birds of Southern Africa*, text by A. Kemp (Johannesburg, 1982); J. Sepp, *Butterflies and Moths*, text by S. McNeill (London, 1978); contrast F. O. Morris, *British Birds*, intr. A. Soper (London, 1985), where the work is presented as a period-piece, and L. J. P. Vieillot, *Songbirds of the Torrid Zone* (Kent, Ohio, 1979). D. J. Carr (ed.), *Sydney Parkinson* (London, 1983); B. Smith, *European Vision and the South Pacific*, 2nd edn. (New Haven, 1985). K. Baynes and F. Pugh, *The Art of the Engineer* (London, 1981); G. Wallis and J. Whitworth, *The American System of Manufactures*, ed. N. Rosenberg (Edinburgh, 1969). E. R. Tufte, *The Visual Display of Quantitative Information* (Cheshire, Conn., 1983).

8 Scientific Culture

On the BAAS, see J. Morrell and A. Thackray, *Gentlemen of Science* (Oxford, 1981), and the texts *Early Correspondence* (London, 1984); R. MacLeod and P. Collins, *The Parliament of Science* (London, 1981). On the 1860 meeting, and the Huxley–Wilberforce encounter, see the papers by J. R. Lucas, *The Historical Journal*, 22 (1979), 313–30, and S. Gilley, *Studies in Church History*, 17 (1981), 325–40. On metropolitan–provincial relations, I. Inkster and J. Morrell, *Metropolis and Province* (London, 1983); G. Kitteringham, 'Science in provincial society: the case of Liverpool', *Annals of Science*, 39 (1982), 329–48; J. Morrell, 'Wissenschaft in Worstedopolis', *BJHS*, 18 (1985), 1–23. On the Royal Institution, M. Berman, *Social Change and Scientific Organization* (London, 1978); D. Chilton and N. G. Coley, 'The laboratories of the Royal Institution', *Ambix*, 27 (1980), 173–203; on the Royal Society, M. P. Crosland, 'Explicit qualifications', *Notes and Records of the Roy. Soc.*, 37 (1983), 167–87; M. B. Hall, *All Scientists Now* (Cambridge, 1984). On the architecture of scientific institutions, see the paper by S. Forgan, 'Context, image and function', *BJHS*,

19 (1986) 89–113. On Mary Somerville, see E. C. Patterson, *Mary Somerville and the Cultivation of Science* (Dordrecht, 1983); on mapping Ireland, J. H. Andrews, *A Paper Landscape* (Oxford, 1975), and G. Herries Davies, *Sheets of Many Colours* (Dublin, 1983); on geological institutions, M. J. S. Rudwick, *The Great Devonian Controversy* (Chicago, 1985); on naval surveys, see A. Day, *The Admiralty Hydrographic Service* (London, 1967); G. S. Ritchie, *The Admiralty Chart* (London, 1967); and my introduction to the reprint of J. F. W. Herschel (ed.), *Admiralty Manual of Scientific Enquiry* (1851) (Folkestone, 1974); I. H. W. Engstrand, *Spanish Scientists in the New World* (Seattle, 1981). On the use of local talent in expeditions, see D. H. Simpson, *Dark Companions* (London, 1975); and the new life of W. Moorcroft, the vet turned explorer, G. Adler, *Beyond Bokhara* (London, 1985). M. Berg, *The Machinery Question and the Making of Political Economy, 1815–1848* (Cambridge, 1980). J. A. Cawood, 'Terrestrial magnetism and the development of international collaboration', *Annals of Science*, 34 (1980) 551–87; M. Heidelberger on Ohm, 'A logical reconstruction of historical change', *Studies in the History and Philosophy of Science*, 11 (1980) 103–21, concerns the change from Humboldtian to exact physics. A. J. Meadows (ed.), *The Development of Science Publishing in Europe* (Amsterdam, 1980). J. Shattock and M. Wolff (eds), *The Victorian Periodical Press* (Toronto, 1982). J. Burton, 'Robert Fitzroy and the Meteorological Office; *BJHS*, forthcoming.

9 Battle of Symbols and Jargon

O. Hannaway, *The Chemists and the Word* (Baltimore, 1975); M. P. Crosland, *Historical Studies in the Language of Chemistry*, 2nd edn. (New York, 1978); W. C. Anderson, *Between the Library and the Laboratory* (Baltimore, 1984); A. M. Duncan, 'Styles of language and modes of chemical thought', *Ambix*, 28 (1981), 83–107; my 'Chemistry and poetic imagery', *Chemistry in Britain*, 19 (1983), 578–82; and *Atoms and Elements* (London, 1970). C. A. Russell (ed.), *Recent Developments in the History of Chemistry* (London, 1985); F. Szabadváry, 'Centuries of research in chemistry', *Periodica Polytechnica, Chemical Engineering*, 28 (1984), 143–55. F. Gettings, *Dictionary of Occult, Hermetic, and Alchemical Sigils* (London, 1981); A. J. Rocke, *Chemical Atomism in the Nineteenth Century* (Columbus, 1984); M. J. Nye, *The Question of the Atom* (Los Angeles, 1984); H. Kragh, 'Julius Thomsen and classical thermochemistry', *BJHS*, 17 (1984), 225–72; R. G. A. Dolby, 'Thermochemistry versus thermodynamics', *History of Science*, 22 (1984), 375–400; K. Hutchinson, 'W. J. M. Rankine and the rise of thermodynamics', *BJHS*, 14 (1981), 1–26; D. Gooding, 'Metaphysics versus measurement', *Annals of Science*, 37 (1980), 1–29. G. N. Cantor and M. J. S. Hodge (ed.), *Conceptions of Ether* (Cambridge, 1981); P. M. Harman, *Energy, Force and Matter* (Cambridge, 1982); G. L. E. Turner, *Nineteenth-century Scientific Instruments* (London, 1983). J. R. Milton, 'The origin and development of the concept of the "laws of nature"', *Arch. Europ. Sociol.*, 22 (1981), 173–95. R. F. Bud and G. K. Roberts, *Science Versus Practice* (Manchester, 1984). T. M. Porter, 'The mathematics of society: variation and error in Quetelet's statistics', *BJHS*, 18 (1985) 51–69. R. Anderson, 'Secondary schools and Scottish society in the nineteenth century', *Past and Present*, 109 (1985), 176–203. H. A. M. Snelders, 'J. H. van't Hoff's research school', *Janus*, 71 (1984), 1–30.

10 The Triumphal Chariot

For a chronology of scientific discoveries, see C. L. Parkinson, *Breakthroughs* (London, 1985); this involves selection, but builds up a picture of very rapid growth in the last century. P. M. Harman is preparing an edition of the letters and papers of Maxwell for Cambridge University Press; his papers on particular topics are being published by S. G. Brush, C. W. F. Everitt and E. Garber (Cambridge, Mass.,1983—). See *Faraday's Diary*, ed. T. Martin, 7 vols., (London, 1932–6), for his techniques and occasionally thoughts put on paper. Mary Hesse, *Forces and Fields* (London, 1961); L. P. Williams, *Michael Faraday* (London, 1965). M. Hoskin, *Stellar Astronomy* (Chalfont St Giles, 1982). On American science, see N. Reingold (ed.), *Science in Nineteenth-century America* (London, 1966); this is a collection of documents. On the problems of organizing museums, see the book by one of Owen's successors: W. H. Flower, *Essays on Museums* (London, 1898). The sumptuous volumes edited by C. Darwin, *The Zoology of the Voyage of HMS* Beagle, were reprinted in facsimile (Wellington, 1980). M. J. S. Rudwick, *The Meaning of Fossils* (London, 1972); J. D. Burchfield, *Lord Kelvin and the Age of the Earth* (New York, 1975). G. L'E. Turner, *Essays on the History of the Microscope* (Oxford, 1980). For a general history of mathematics, K. Kline, *Mathematical Thought from Ancient to Modern Times* (New York, 1972). On the lighter side, A. Sutton, *The Victorian World of Science* (Bristol, 1985). On education: W. H. Brock, 'From Liebig to Nuffield. A bibliography of the history of science education, 1839–1974', *Studies in Science Education*, 21 (1975), 67–99; M. Sanderson (ed.), *The Universities in the Nineteenth Century* (London, 1975); J. Roach, *Public Examinations in England, 1850–1900* (Cambridge, 1971); R. MacLeod (ed.), *Days of Judgement* (Driffield, 1982); P. Allen, *The Cambridge Apostles* (Cambridge, 1978); H. W. Becher, 'Voluntary science at Cambridge', *BJHS* 19 (1986), 57–88; and compare D. K. Müller, 'The qualifications crisis and school reform in late nineteenth-century Germany', *History of Education*, 19 (1980), 315–31. E. R. Robson, *School Architecture* (1874) intr. M. Seaborne, (Leicester, 1972).

11 Wrestling with the Unknown

W. H. Brock and J. MacLeod, 'The scientists' declaration', *BJHS*, 9 (1976), 39–66. A. Gauld, *The Founders of Psychical Research* (London, 1968); A. Gregory, 'The anatomy of a fraud', *Annals of Science*, 34 (1977), 449–549; on separating real and pseudo-science, see I. Lakatos, *The Methodology of Scientific Research Programmes* (Cambridge, 1978), pp. 1–7. On spectroscopy, see M. A. Sutton, 'Spectroscopy and the chemists', *Ambix*, 23 (1976), 16–26, and the papers of F. A. J. L. James, 'Spark spectra' and 'Line spectra', *Ambix*, 30 (1983), 137–62, and 32 (1985), 53–70. On Crookes, S. B. Sinclair, 'Crookes and radioactivity', *Ambix*, 32 (1985), 15–31; R. K. DeKosky, 'William Crookes and the quest for absolute vacuum in the 1870s', *Annals of Science*, 40 (1983), 1–18, and 'George Gabriel Stokes, Arthur Smithells and the origin of spectra in flames', *Ambix*, 27 (1980), 103–23. Tait's mathematical work was based on that of Hamilton; see T. L. Hankins, *Sir William Rowan Hamilton* (Baltimore, 1980). On Lockyer and *Nature*, see A. J. Meadows, *Science and Controversy* (London, 1972); on Röntgen, W. R. Nitske, *Wilhelm Conrad Röntgen* (Tucson, 1971); on Galton and evolutionary anthropology, R. E. Fancher, 'Francis Galton's African ethnology', *BJHS*, 16 (1983), 67–79.

12 Science Comes of Age

R. McCormach, *Night Thoughts of a Classical Physicist* (Cambridge, Mass., 1982), is a tour-de-force on the period; and on Einstein, see A. Pais, *Subtle is the Lord* (Oxford, 1982). J. Hendry (ed.), *Cambridge Physics in the Thirties* (Bristol, 1984), and 'The development of attitudes to wave-particle duality', *Annals of Science*, 37 (1980), 59–79. On the export of science, W. H. Brock, 'The Japanese connexion', *BJHS*, 14 (1981), 227–43; M. Gorman, 'Sir William O'Shaughnessy, pioneer chemical educator in India', *Ambix*, 30 (1983), 107–116; A. R. Choudhuri, 'Practising Western science outside the West', *Social Studies in Science*, 15 (1985), 475–505. On scientism and mankind, S. J. Gould, *The Mismeasure of Man* (Harmondsworth, 1984). On sources for the history of science, see my *Sources* (Cambridge, 1975); new Japanese edition, trans. H. Kasiwagi (Tokyo, 1984); and S. A. Jayawardene, *Reference Books for the Historian of Science* (London, 1982). On modern science generally, B. Barnes, *About Science* (Oxford, 1984); J. Marks, *Science and the Modern World* (London, 1983); N. Clark, *The Political Economy of Science and Technology* (Oxford, 1985).

Notes

Chapter 1: The Age of Science?

1 Jane Marcet, *Conversations on Chemistry* (London, 1806).
2 Arthur Schuster, *The Progress of Physics* (Cambridge, 1911), p. 25.

Chapter 2: Rivalry with the French

1 M. P. Crosland, *The Society of Arcueil* (London, 1967).
2 C. C. Gillispie, *The Montgolfier Brothers* (Princeton, 1983).
3 D. Outram (ed.), *The Letters of Georges Cuvier: a Summary Calendar* (Chalfont St Giles, 1980); and D. Outram, *Georges Cuvier* (Manchester, 1984).
4 A. D. C. Simpson (ed.), *Joseph Black* (Edinburgh, 1982).
5 John Robison, *Proofs of a Conspiracy against all the Religions and Governments of Europe* (Edinburgh, 1797), p. 446.
6 J. W. Yolton, *Thinking Matter* (Oxford, 1984); review of Yolton by G. Cantor, *History of Science*, 23 (1985), 201–6; D. Knight, *The Transcendental Part of Chemistry* (Folkestone, 1978).
7 L. J. Jordanova, *Lamarck* (Oxford, 1984).
8 R. W. Burckhardt, *Spirit of System* (Cambridge, Mass., 1977).
9 R. Yeo, 'William Whewell', *Annals of Science*, 36 (1979), 493–516; and 'Baconianism', *History of Science*, 23 (1985), 251–98.
10 Francis Bacon, *The Advancement of Learning* (London, 1605), I: 1 and 3.
11 S. Carnot, *Réflexions*, ed. R. Fox (Paris, 1978).
12 S. Forgan (ed.), *Science and the Sons of Genius* (London, 1980).
13 M. P. Crosland, *Gay-Lussac* (Cambridge, 1978), ch. 4.
14 A. D. C. Simpson, *Joseph Black*; M. Foucault, *The Birth of the*

Clinic, trans. A. M. Sheridan Smith (London, 1973); J. Woodward, *To Do the Sick No Harm* (London, 1974).

15 H. W. Becher, 'Voluntary science at Cambridge', *BJHS*, 19 (1986), 57–87.
16 D. Gooding and F. James (eds), *Faraday Rediscovered* (London, 1985), pp. 33ff.
17 L. J. Klosterman, 'A research school of chemistry in the nineteenth century', *Annals of Science*, 42 (1985), 1–40.
18 D. P. Miller, 'Between hostile camps', *BJHS*, 16 (1983), 1–47.
19 R. W. Home, 'Poisson's memoirs on electricity', *BJHS*, 16 (1983), 239–59.
20 M. P. Crosland, *The Society of Arceuil*, ch. 6.
21 G. Cantor, *Optics After Newton* (Manchester, 1983), pp. 129ff.
22 A. J. Rocke, *Chemical Atomism in the Nineteenth Century* (Columbus, 1984).
23 M. Deacon, *Scientists and the Sea* (London, 1971), pp. 175ff.
24 N. Baudin, *Journal*, trans. C. Cornell (Adelaide, 1974).
25 D. J. Mabberly, *Jupiter Botanicus* (Braunschweig, 1985).
26 G. Vevers, *London's Zoo* (London, 1976).

Chapter 3: Wrestling with God

1 C. A. Russell, *Cross-currents* (London, 1985).
2 M. Hunter, *Science and Society in Restoration England* (Cambridge, 1981).
3 J. North and J. Roche (eds), *The Light of Nature* (Dordrecht, 1985), pp. 83ff.
4 D. Knight, *Natural Science Books in English, 1600–1900* (London, 1972), ch. 3.
5 A. W. Colver and J. V. Price (eds), *Hume on Religion* (Oxford, 1976).
6 L. J. Jordanova and R. Porter (eds), *Images of the Earth* (Chalfont St Giles, 1978).
7 M. J. S. Rudwick, *The Meaning of Fossils* (London, 1972).
8 D. Outram (ed.), *The Letters of Georges Cuvier: a summary calendar* (Chalfont St Giles, 1980); D. Outram, *Georges Cuvier* (Manchester, 1984).
9 N. A. Rupke, *The Great Chain of History* (Oxford, 1983), pp. 130ff.
10 H. Davy, *Works* (1839–40), vol. 7, pp. 33ff.
11 D. Knight, *Natural Science Books*, ch. 3.
12 R. Yeo, 'William Whewell'.

13　W. H. Brock, *From Protyle to Proton* (Bristol, 1985).

14　P. Butler (ed.), *Pusey Rediscovered* (London, 1983), esp. ch. 1.

15　J. H. Newman, *A Grammar of Assent*, intr. N. Lash (Notre Dame, 1979), esp. ch. 4.

16　T. H. Levere, *Poetry Realized in Nature* (Cambridge, 1981).

17　See chapter 2, note 8; and R. Yeo, 'Science and intellectual authority', *Victorian Studies*, 28 (1984), 5–31.

18　T. M. Porter, 'The mathematics of society: variation and error in Quetelet's statistics', *BJHS*, 18 (1985), 51–69.

Chapter 4: The German Challenge

1　A. R. Hall, *Philosophers at War* (Cambridge, 1980).

2　O. Hannaway, *The Chemists and the Word* (Baltimore, 1975).

3　W. C. Anderson, *Between the Library and the Laboratory* (Baltimore, 1984).

4　D. Knight, 'Chemistry and poetic imagery', *Chemistry in Britain*, 19 (1983), 578–82.

5　T. Lenoir, *The Strategy of Life* (Dordrecht, 1982).

6　G. L'E. Turner, *Essays on the History of the Microscope* (Oxford, 1980).

7　F. Gregory, 'Romantic Kantianism and the end of the Newtonian dream in chemistry', *Archives internationales d'histoire des sciences*, 34 (1984), 108–23.

8　G. W. F. Hegel, *Philosophy of Nature*, ed. and trans. M. J. Petry (London, 1970); M. J. Petry and R. Horstmann (eds), *Hegels Philosophie der Natur* (Stuttgart, 1986).

9　N. Reingold (ed.), *The Papers of Joseph Henry* (Washington, 1972—); T. H. Levere and R. Jarrell, *A Curious Field-book* (Toronto, 1974); S. G. Kohlstedt, *The Formation of the American Scientific Community* (Urbana, 1976); A. P. Molella, 'At the edge of science', *Annals of Science*, 41 (1984), 445–61; L. Owens, 'Pure and sound government', *Isis*, 76 (1985), 182–94.

10　P. R. Sweet, *Wilhelm von Humboldt* (Columbus, 1978–80); C. E. McClelland, *State, Society and University in Germany 1700–1914* (Cambridge, 1980) pt. 2.

11　R. Anderson, 'Secondary schools and Scottish society in the nineteenth century', *Past and Present*, 109 (1985), 176–203; D. K. Müller, 'The qualifications crisis and school reform in late nineteenth-century Germany', *History of Education*, 9 (1980), 315–31; E. R. Robson, *School Architecture* (1874), intr. M. Seaborne (Leicester, 1972), ch. 6.

12 R. Paul, 'German academic science and the mandarin ethos', *BJHS*, 17 (1984), 1–29.

13 K. Hufbaur, *The Formation of the German Chemical Community, 1720–1795* (Berkeley, 1982).

14 W. H. Brock, 'Liebigiana', *History of Science*, 19 (1981), 201–18.

15 A. Desmond, *Archetypes and Ancestors* (London, 1982).

16 J. C. Greene, *Science, Ideology and World-view* (Berkeley, 1981).

17 S. Hahnemann, *Organon of Medicine*, trans. J. Künzli, A. Nandi and P. Pendleton (London, 1983).

18 J. A. Cawood, 'Terrestrial magnetism and the development of international collaboration', *Annals of Science*, 34 (1980), 551–87; M. Heidelberger, 'A logical reconstruction of historical change', *Studies in the History and Philosophy of Science*, 11 (1980), 103–21.

19 T. H. Levere, *Poetry Realized*, pp. 36ff. An important vehicle for German philosophy was J. D. Morell *The speculative Philosophy of Europe in the 19th Century* (2nd edn., London, 1847).

20 T. S. Kuhn, *The Essential Tension* (Chicago, 1977).

21 W. H. Brock and A. J. Meadows, *The Lamp of Learning* (London, 1984), ch. 4.

22 E. Du Bois-Reymond, P. Diepgen and P. C. Cranefield (eds), *Two Great Scientists*, trans. S. Lichtner-Ayéd (Baltimore, 1982).

23 G. L. Geison, *Michael Foster and the Cambridge School of Physiology* (Princeton, 1978); R. E. Kohler, *From Medical Chemistry to Biochemistry* (Cambridge, 1982).

24 E. Wilson, *Adorned in Dreams* (London, 1985).

Chapter 5: Arguing with Sceptics

1 G. Tytler, *Physiognomy in the European Novel* (Princeton, 1982).

2 G. N. Cantor, 'The Edinburgh phrenological debate' and S. Shapin, 'Phrenological knowledge and social structure', *Annals of Science*, 32 (1975), 195–256.

3 S. J. Gould, *The Mismeasure of Man* (Harmondsworth, 1984), ch. 3.

4 D. Knight, *Transcendental Part of Chemistry*, ch. 3.

5 J. H. Brooke, 'Natural theology and the plurality of worlds', *Annals of Science*, 34 (1977), 221–86; S. J. Dick, *Plurality of Worlds* (Cambridge, 1982).

6 R. Yeo, 'William Whewell'.

7 A. D. Morrison-Low and J. R. R. Christie (eds), *Martyr of Science: David Brewster* (Edinburgh, 1984).

8 R. W. Burckhardt, *Spirit of System*; L. J. Jordanova, *Lamarck*.
9 M. B. Ogilvie, 'Robert Chambers and the nebular hypothesis', *BJHS*, 8 (1975), 314–32; J. H. Brooke, 'Richard Owen, William Whewell, and the *Vestiges*', *BJHS*, 10 (1977), 132–45.
10 M. A. Sutton, 'Spectroscopy and the chemists', *Ambix*, 23 (1976), 16–26.
11 G. Cantor, *Optics after Newton* (Manchester, 1983), ch. 6.
12 P. M. Harman, *Energy, Force and Matter* (Cambridge, 1982).
13 G. N. Cantor and M. J. S. Hodge (eds), *Conceptions of Ether* (Cambridge, 1981).
14 D. Knight, *Atoms and Elements*, 2nd edn (London, 1970), ch. 4.
15 F. Gregory, 'Romantic Kantianism'; C. Smith, 'Natural philosophy and thermodynamics', *BJHS*, 9 (1976), 293–319.
16 T. Lenoir, *The Strategy of Life* (Dordrecht, 1982).

Chapter 6: Debate about Animals

1 D. Knight, *Ordering the World* (London, 1981); M. P. Winsor, *Starfish, Jellyfish, and the Order of Life* (New Haven, 1976).
2 M. S. Merian, *The Wonderful Transformations of Caterpillars*, ed. W. T. Stearn (London, 1978).
3 I. S. Majneep and R. Bulmer, *Birds of my Kalam Country* (Oxford, 1977).
4 W. T. Stearn, *Botanical Latin*, 2nd edn (Newton Abbot, 1973).
5 P. R. Sloan, 'Buffon, German biology, and the historical interpretation of biological species', *BJHS*, 12 (1979), 109–53.
6 D. Outram, *Georges Cuvier*.
7 L. J. Jordanova, *Lamarck*.
8 M. P. Winsor, *Starfish*.
9 A. Desmond, *Archetypes*.
10 D. Knight, 'William Swainson, author and illustrator', *Archives of Natural History*, forthcoming 1986; and 'William Swainson; types, circles and affinities', in J. North and J. Roche (eds), *The Light of Nature* (Dordrecht, 1985).
11 S. Hyman, *Edward Lear's Birds* (London, 1985).
12 S. Forgan, 'Context, image and function', *BJHS*, 19 (1986), 89–113.
13 J. Morrell and A. Thackray, *Gentlemen of Science* (Oxford, 1981).
14 G. Wilson and A. Geikie, *Edward Forbes* (Edinburgh, 1861).
15 D. E. Allen, *The Naturalist in Britain* (London, 1976).
16 S. Sheets-Pyenson, 'Geological communication', *Bulletin of the British Museum (Nat. Hist.), Hist. Series*, 10 (1982), 179–226;

J. Secord, 'John W. Salter', in A. Wheeler and J. H. Price (eds), *From Linnaeus to Darwin* (London, 1985), pp. 61–75.

17 D. J. Mabberly, *Jupiter Botanicus* (Braunschweig, 1985), part 3.

18 D. Ripley, *The Sacred Grove* (London, 1970); R. Desmond, *The India Museum* (London, 1982); A. E. Gunther, *A Century of Zoology at the British Museum* (London, 1975); M. Girouard, *Alfred Waterhouse and the Natural History Museum* (London, 1981).

19 A. Desmond, *Archetypes*. Cf. M. Midgley, *Evolution as a Religion* (London, 1985).

20 D. R. Oldroyd, *Darwinian Impacts* (Milton Keynes, 1980).

21 N. Reingold (ed.), *Science in Nineteenth-century America* (London, 1966), pp. 29ff, 162ff.

22 J. R. Moore, *The Post-Darwinian Controversies* (Cambridge, 1979).

23 P. J. Bowler, *The Eclipse of Darwinism* (Baltimore, 1983).

24 G. L. Geison, *Michael Foster*; R. E. Kohler, *From Medical Chemistry*.

Chapter 7: Discourse in Pictures

1 A. Ellenius (ed.), *The Natural Sciences and the Arts*, *Acta Universitatis Upsaliensis*, 22 (Uppsala, 1985).

2 R. B. Freeman, *British Natural History Books* (Folkestone, 1980); D. Knight, *Zoological Illustration* (Folkestone, 1977).

3 M. J. S. Rudwick, 'A visual language for geology', *History of Science*, 14 (1976), 149–95.

4 S. Hyman, *Edward Lear's Birds*.

5 G. C. Saur, *John Gould* (Melbourne, 1982).

6 W. T. Stearn (ed.), *The Australian Flower Paintings of Ferdinand Bauer* (London, 1976).

7 L. J. P. Vieillot, *Songbirds of the Torrid Zone* (Kent, Ohio, 1979).

8 J. R. Lucas, 'Wilberforce and Huxley: a legendary encounter', *The Historical Journal*, 22 (1979), 313–30; S. Gilley, 'The Huxley–Wilberforce Debate', *Studies in Church History*, 17 (1981), 325–40.

9 A. Rutgers, *Birds of New Guinea*, illus. J. Gould (London, 1970); G. E. Lodge, *Unpublished Bird Paintings*, text by C. A. Fleming (London, 1983); C. G. Finch-Davies, *The Birds of Southern Africa*, text by A. Kemp (Johannesburg, 1982); J. Sepp, *Butterflies and Moths*, text by S. McNeill (London, 1978); contrast F. O. Morris, *British Birds*, intr. A. Soper (London,

1985); L. J. P. Vieillot, *Songbirds*; where the emphasis is on context, not modernity.

10 M. J. S. Rudwick, *Meaning of Fossils*, ch. 3.
11 N. A. Rupke, *The Great Chain of History*.
12 A. Desmond, *Archetypes and Ancestors*, ch. 4.
13 M. Pollock (ed.), *Common Denominators in Art and Science* (Aberdeen, 1983), pp. 30ff.
14 D. J. Carr, *Sydney Parkinson* (London, 1983).
15 B. Smith, *European Vision and the South Pacific*, 2nd edn (New Haven, 1985).
16 I. H. W. Engstrand, *Spanish Scientists in the New World* (Seattle, 1981).
17 A. Ellenius (ed.), *Natural Sciences and Arts*, 121f, 150ff.
18 T. Keulemans and J. Coldeway, *Feathers to Brush* (Melbourne, 1982).
19 K. Baynes and F. Pugh, *The Art of the Engineer* (London, 1981).
20 P. Lervig, 'Sadi Carnot and the steam engine', *BJHS*, 18 (1985), 147–96.
21 C. C. Gillispie (ed.), *A Diderot Pictorial Encyclopedia of Trades and Industry* (New York, 1959); contrast R. Fox, 'Science, industry and the social order in Mulhouse, 1798–1871', *BJHS*, 17 (1984), 127–68.
22 G. Wallis and J. Whitworth, *The American System of Manufactures*, ed. N. Rosenberg (Edinburgh, 1969); B. Sinclair, *Philadelphia's Philosopher Mechanics* (Baltimore, 1974); M. R. Smith, *Harpers Ferry Armory and the New Technology* (Ithaca, 1977).

Chapter 8: Scientific Culture

1 J. Morrell and A. Thackray, *Gentlemen of Science* and *Early Correspondence* (London, 1984); R. MacLeod and P. Collins, *The Parliament of Science* (London, 1981).
2 J. R. Lucas, 'Wilberforce and Huxley'; S. Gilley, 'Huxley–Wilberforce Debate'.
3 S. Forgan, 'Context, image and function'.
4 J. Burton, 'Robert FitzRoy and the Meteorological Office', *BJHS*, forthcoming 1986.
5 J. A. Cawood, 'Terrestrial magnetism'; M. Heidelberger, 'Logical reconstruction of historical change'.
6 J. R. Lucas, 'Wilberforce and Huxley'.
7 I. Inkster and J. Morrell (eds), *Metropolis and Province* (London, 1983); G. Kitteringham, 'Science in provincial society: the case

of Liverpool', *Annals of Science*, 39 (1982), 329–48; J. Morrell, 'Wissenschaft in Worstdopolis', *BJHS*, 18 (1985), 1–23.

8 M. P. Crosland, *Society of Arceuil*.

9 S. G. Kohlstedt, *American Scientific Community*; L. P Williams, *Album of Science* (New York, 1978), esp. 19f.

10 S. Sheets-Pyenson, 'Geological communication'; J. Secord, 'John W. Salter'. W. G. Adams and J. W. L. Glaisher, *The Scientific Papers of J. C. Adams* (Cambridge, 1896) XV ff.

11 J. Morrell and A. Thackray, *Gentlemen of Science*.

12 D. E. Allen, *The Naturalist in Britain*.

13 M. Berman, *Social Change and Scientific Organization* (London, 1978).

14 A. J. Meadows (ed.), *The Development of Science Publishing in Europe* (Amsterdam, 1980); J. Shattock and M. Wolff (eds), *The Victorian Periodical Press* (Toronto, 1982).

15 M. P. Crosland, 'Explicit qualifications', *Notes and Records of the Roy. Soc.*, 37 (1983), 167–87; M. B. Hall, *All Scientists Now* (Cambridge, 1984).

16 E. C. Patterson, *Mary Somerville and the Cultivation of Science* (Dordrecht, 1983).

17 E. R. Robson, *School Architecture* (1874), intr. M. Seaborne (Leicester, 1972).

18 D. Chilton and N. G. Coley, 'The laboratories of the Royal Institution', *Ambix*, 27 (1980), 173–203; S. Forgan, 'Context, Image and Function'.

19 J. C. Greene, *Science, Ideology*.

20 W. H. Flower, *Essays on Museums* (London, 1898).
   M. J. S. Rudwick, *The Great Devonian Controversy* (Chicago, 1985).

22 I. H. W. Engstrand, *Spanish Scientists*; J. H. Ostrom and J. S. McIntosh, *Marsh's Dinosaurs* (New Haven, 1966).

Chapter   9: The Battle of Symbols and Jargon

1 O. Hannaway, *The Chemists and the Word* (Baltimore, 1975).

2 F. Gettings, *Dictionary of Occult, Hermetic, and Alchemical Sigils* (London, 1981).

3 M. P. Crosland, *Historical Studies in the Language of Chemistry*, 2nd edn (New York, 1978); W. C. Anderson, *Between the Library and the Laboratory* (Baltimore, 1984).

4 A. M. Duncan, 'Styles of language and modes of chemical thought', *Ambix* 28 (1981), 83–107; D. Knight, 'Chemistry and poetic imagery'.

5   A. J. Rocke, *Chemical Atomism in the Nineteenth Century* (Columbus, 1984); M. J. Nye, *The Question of the Atom* (Los Angeles, 1984).

6   D. Knight, *Transcendental Part of Chemistry*.

7   D. Knight, *Atoms and Elements*, ch. 2.

8   G. W. F. Hegel, *Philosophy of Nature*; M. J. Petry and R. Horstmann, *Hegels Philosophie der Natur*.

9   M. A. Sutton, 'Spectroscopy'.

10 H. Kragh, 'Julius Thomsen and classical thermochemistry', *BJHS*, 17 (1984), 255–72; R. G. A. Dolby, 'Thermochemistry versus thermodynamics', *History of Science*, 22 (1984), 375–400; K. Hutchinson, 'W. J. M. Rankine'; G. N. Cantor and M. J. S. Hodge, *Conceptions of Ether*; D. Gooding, 'Metaphysics versus measurement', *Annals of Science*, 37 (1980), 1–29; P. M. Harman, *Energy*.

11 T. S. Kuhn, *The Essential Tension* (Chicago, 1977).

12 R. W. Home, 'Poisson's memoirs on electricity', *BJHS*, 16 (1983), 239–59.

13 J. R. Milton, 'The origin and development of the concept of the "laws of nature"', *Arch. Europ. Sociol.*, 22 (1981), 173–95.

14 H. Davy, *Works* (London, 1839–40), vol. 7 pp. 93ff.

15 S. Carnot, *Réflexions*.

16 W. D. Niven (ed.), *The Scientific Papers of James Clerk Maxwell* (Cambridge, 1890), vol. 1, pp. 377ff; S. G. Brush, C. W. F. Everitt and E. Garber (eds), *Maxwell on Saturns Rings* (1983); *Maxwell on Molecules and Gases* (1986) (Cambridge, Mass. continuing). C. G. Knott, *Life and Scientific Work of P. G. Tait* (Edinburgh, 1911) pp. 213ff.

17 T. M. Porter, 'Mathematics of society'.

18 R. F. Bud and G. K. Roberts, *Science Versus Practice* (Manchester, 1984), pp. 19ff.

19 I. Inkster and J. Morrell (ed.), *Metropolis and Province*.

20 R. F. Bud and G. K. Roberts, *Science Versus Practice*, pp. 71ff.

Chapter 10: The Triumphal Chariot

1   C. L. Parkinson, *Breakthroughs* (London, 1985).

2   G. L'E. Turner, *Nineteenth-century Scientific Instruments* (London, 1983).

3   E. Du Bois-Reymond, P. Diepgen and P. C. Cranefield, *Two Great Scientists*.

4   S. Forgan, 'Context, image and function'.
5   D. Gooding and F. James (eds), *Faraday Rediscovered*.
6   T. Martin (ed.), *Faraday's Diary*, 7 vols (London, 1932–6).
7   L. P. Williams, *Michael Faraday* (London, 1965), pp. 73ff, 408ff.
8   G. M. Cantor and M. J. S. Hodge, *Conceptions of Ether*.
9   M. Faraday, *Experimental Researches*, vol. 3 (London, 1855), p. 450.
10 M. Hesse, *Forces and Fields* (London, 1961).
11 N. Reingold, *Papers of Joseph Henry*, vol. 1, pp. 375, 436ff.
12 P. M. Harman, *Energy*.
13 L. Campbell and W. Garnett, *The Life of J. C. Maxwell* (Cambridge, 1882).
14 J. A. Cawood, 'Terrestrial magnetism'; M. Heidelberger, 'Logical reconstruction of historical change'.
15 R. K. DeKosky, 'George Gabriel Stokes, Arthur Smithells and the origin of spectra in flames', *Ambix*, 27 (1980), 103–23.
16 M. A. Sutton, 'Spectroscopy'; F. A. J. L. James, 'Spark spectra', *Ambix*, 30 (1983), 137–62, and 'Line spectra', *Ambix*, 32 (1985), 53–70.
17 M. Hoskin, *Stellar Astronomy* (Chalfont St. Giles, 1982).
18 M. Alic, *Hypatia's Legacy* (London, 1983).
19 A. J. Meadows, *Science and Controversy* (London, 1972), pp. 58ff.
20 G. L'E. Turner, *Nineteenth-century Scientific Instruments*.
21 W. C. Williamson, *Reminiscences* (1896, reprint Manchester, 1985), p. 66.
22 D. J. Mabberly, *Jupiter Botanicus*, part III; A. E. Gunther, *A Century of Zoology*.
23 M. Girouard, *Alfred Waterhouse*.
24 W. H. Flower, *Essays on Museums*.
25 C. Darwin (ed.), *The Zoology of the Voyage of HMS Beagle* (facsimile reprint, Wellington, 1980), vol. 2.
26 N. Reingold, *Science in Nineteenth-century America*, pp. 127ff.
27 D. R. Oldroyd, *Darwinian Impacts*.
28 C. Lyell, *Scientific Journals on the Species Question*, ed. L. G. Wilson (New Haven, 1970).
29 J. D. Burchfield, *Lord Kelvin and the Age of the Earth* (New York, 1975).
30 P. J. Bowler, *The Eclipse of Darwinism*, ch. 1.
31 L. J. Jordanova, *Lamarck*.
32 G. L'E. Turner, *History of the Microscope*.
33 C. A. Russell (ed.), *Recent Developments in the History of Chemistry* (London, 1985).
34 T. M. Porter, 'Mathematics of society'.

35   M. Kline, *Mathematical Thought from Ancient to Modern Times* (New York, 1972).

36   T. L. Hankins, *Sir William Rowan Hamilton* (Baltimore, 1980).

37   R. Paul, 'German academic science'; D. K. Müller, 'The qualifications crisis'. J. A. Johnston 'Academic chemistry in Imperial Germany', *Isis*, 76 (1985) 500–24.

38   R. McCormach, *Night Thoughts of a Classical Physicist* (Cambridge, Mass., 1982); T. Zeldin, *France, 1848–1975*, vol. 2 (Oxford, 1977), ch. 12.

39   P. Allen, *The Cambridge Apostles* (Cambridge, 1978); H. W. Becher, 'Voluntary science at Cambridge'.

40   G. L. Geison, *Michael Foster*.

41   W. H. Brock, 'From Liebig to Nuffield: a bibliography of the history of science education, 1839–1974', *Studies in Science Education*, 21 (1975), 67–99.

42   L. A. Farrell, 'The history of eugenics', D. MacKenzie, 'Karl Pearson and the professional middle class', R. Love, 'Alice in eugenics-land', J. R. Searle, 'Eugenics and politics in Britain', *Annals of Science*, 36 (1979), 111–69.

Chapter   11: Wrestling with the Unknown

1   O. Chadwick, *The Victorian Church* (London, 1966–70), vol. 2, ch. 1.

2   W. H. Brock and J. MacLeod, 'The scientists' declaration', *BJHS*, 9 (1976), 39–66.

3   S. Forgan (ed.), *Science and the Sons of Genius* (London, 1980), pp. 38ff.

4   W. K. Clifford, *Lectures and Essays* (London, 1879), vol. 1, pp. 228ff.

5   A. Gauld, *The Founders of Psychical Research* (London, 1968).

6   A. Gregory, 'The anatomy of a fraud', *Annals of Science*, 34 (1977), 449–549.

7   I. Latakos, *The Methodology of Scientific Research Programmes* (Cambridge, 1978), pp. 1–7.

8   R. K. DeKosky, 'William Crookes and the quest for absolute vacuum in the 1870s', *Annals of Science*, 40 (1983), 1–18.

9   J. R. Durant, 'Scientific naturalism and social reform in the thought of A. R. Wallace', *BJHS*, 12 (1979), 31–58.

10   A. J. Meadows, *Science and Controversy*, pp. 25ff.

11   J. J. Thomson, *Recollections and Reflections* (London, 1936), pp. 377ff.

12   E. E. F. d'Albe, *Sir William Crookes* (London, 1923), p. 69.

13   A. J. Meadows, *Development of Science Publishing*, pp. 95ff.
14   M. Shattock and M. Wolff, *Victorian Periodical Press*, part I.
15   E. E. F. d'Albe, *Crookes*, p. 187.
16   R. K. DeKozky, 'William Crookes'.
17   M. A. Sutton, 'Spectroscopy'; F. A. J. L. James, ' "Spark spectra" and "Line spectra" '.
18   W. R. Nitske, *Wilhelm Conrad Röntgen* (Tucson, 1971).
19   J. C. Greene, *Science, Ideology*.
20   R. E. Fancher, 'Francis Galton's African ethnology', *BJHS*, 16 (1983), 67–79.

Chapter   12: Science Comes of Age

1   W. H. Brock, 'The Japanese Connexion', *BJHS*, 14 (1981), 227–43.
2   W. H. Brock (ed.), *The Atomic Debates* (Leicester, 1967), pp. 22–3.
3   A. Pais, *Subtle is the Lord* (Oxford, 1982).
4   J. Hendry, 'The development of attitudes to wave-particle duality', *Annals of Science*, 37 (1980), 59–79.
5   F. N. Cantor and M. J. S. Hodge, *Conceptions of Ether*.
6   R. McCormach, *Night Thoughts*.
7   E. Wilson, *Adorned in Dreams*.
8   I. H. W. Engstrand, *Spanish Scientists*; D. H. Simpson, *Dark Companions* (London, 1975); G. Adler, *Beyond Bokhara* (London, 1985).
9   S. J. Gould, *Mismeasure of Man*.
10   D. R. Oldroyd, *Darwinian Impacts*; J. C. Greene, *Science, Ideology*.
11   L. A. Farrell, 'History of eugenics'; D. MacKenzie, 'Karl Pearson'; R. Love, 'Alice in eugenics-land'; J. R. Searle, 'Eugenics and politics'.
12   J. B. Mozley, *Essays* (London, 1878), vol. 2, p. 338.
13   Ibid., p. 78.

Index

Index compiled by Ann Barham